LESSONS LEARNED: CONNECTING NEW BOILERS TO OLD PIPES

Things you should know when replacing old commercial hydronic boilers

———————

LESSONS LEARNED: CONNECTING NEW BOILERS TO OLD PIPES ver1.01

Copyright © 2014 by Ray Wohlfarth

If you have any questions or comments, please send them to

Ray Wohlfarth
www.FireIceHeat.com
834 Kerry Hill Drive Pittsburgh, PA 15234
Tel 412-343-4110 Fax 412-343-4115
ray@fireiceheat.com

This book is a accumulation of ideas that I have seen and or used in my 30 plus years inside a boiler room. It is not a design manual. This does not take the place of the boiler manufacturers written instructions, engineering, or code issues that may be in force in your locale. Please follow the boiler manufacturers written instructions that are included in the boiler installation manual. This does not take the place of a properly designed system from an experienced designer. Thank you for choosing to purchase and read this book. If you find an error, please let me know so that I could change it for the next issue.

Dedication to the following:

My family, Sheila, Jon, Ryan, Abby, and Conor

My mentor and friend, Dan Holohan, for his support and guidance.

My expert proof readers Jon Wohlfarth, Ryan Wohlfarth, Chuck Ray, Greg Hahn, Ken Womack,

I would also like to acknowledge and thank the following individuals and organizations for helping me in my research for this book.

Boiler Buddy

Ron Lukcic with Chemway

Joel Grabe for helping me while writing this book.

Neutrasafe Corp

Rel-Tek, Inc

Triad Boiler Systems, Inc.

Mr. Harald Prell with Viessman

Mr. Robert Bean

Mr. Joe Fiedrich

A note from the author: Since this book is about hydronic boilers, I purposely made the book 212 pages long, just to make you think. Cool huh? Unless, of course you use Celsius then this is not significant.

Are Boilers Extinct?

"So, what do you do?" the young man asked.

"I sell boilers." I replied

"Really? I didn't think anyone used boilers anymore." he said

"Jerk." I thought to myself.

It sometimes seems that boilers have lost their luster in the industry. They are being overshadowed by sexier systems like geothermal, variable refrigerant flow, and ductless mini split systems as well as older technology like packaged rooftop units. While these systems have a growing footprint in the industry, I believe that boilers still provide a cost effective and safe way to heat a building. Consider these advantages:

Safety While you would never use a high pressure steam boiler to provide space heat, many never give a second thought to installing copper tubes, filled with a toxic gas, R410A, throughout a building for a ductless mini split or variable refrigerant flow system. These systems operate at pressures more than twice that of high pressure steam boilers. In addition, R410A displaces oxygen and is known to affect the central nervous system and cause cardiac arrhythmia. A leaking hydronic system will cause a wet floor. I have installed and serviced many of these myself.

Leakage According to the US Department of Energy, ducted systems can lose up to 40% of the energy generated through duct leaks. Air leaks are difficult to find and repair. Refrigerant leaks could send the refrigerant into the living space. If a hydronic system leaks, cleanup is done with a mop.

Efficiency "How come rooftop units never have snow on them?" my friend Amos used to ask and he got me thinking. In my 30 plus years of servicing commercial heating and cooling systems, I have never seen a packaged rooftop unit with snow on top of the heating section. If you consider that it requires between 250-450 Btu's per square foot to melt snow, imagine how much energy is lost through the top and sides of a packaged rooftop unit compared to a hydronic heating system. In addition, a fan requires 10 times the energy of an ECM pump to transport Btu's and water transfers heat 3,500 times more efficiently than air, according to heating expert John Siegenthaler, P.E.

Smaller Footprint A 1" pipe can transport as many Btus as a 14" round or a 10" x 18" duct. This makes them cost effective for the distribution of Btu's.

Healthier If someone has a virus or cold inside a building with an air handling system, those germs are sucked up into the return air duct and shared with everyone. Hydronic systems have virtually no cross contamination

Comfort In my opinion, hydronic systems offers unparalleled comfort without drafty or noisy rooms.

Table of Contents

The Power Struggle

Have you ever found yourself in the middle of an internal power struggle? That is exactly what happened to me on a boiler project for a local university. The director of maintenance for the university called and wanted to use my boilers to replace a leaking old cast iron boiler for one of the buildings. Unknown to the director of maintenance, the finance director hired an engineer, who happened to be a relative, to design the new heating system. Obviously, they do not communicate well. I called the engineer and asked to see him about the project.

"I am familiar with your boiler and they will not work for this project. We want the most efficient boiler we can get." the engineer informed me.

"I have both standard efficiency as well as condensing boilers." I replied

"Their boilers are 99% efficient and yours are only 96% efficient. Every little bit counts." he answered in a most condescending way.

When I looked at his design, I shook my head and said with respect, "You are connecting the new boiler to a system that was designed for 180^0F and the system will be lucky to get about 85% efficiency. When using the condensing boiler to provide space heat, heat the pool and the domestic hot water, it will not condense." It went downhill after that. The engineer pointed his finger at the boiler brochure and said "See, it says 99% efficient. We are done." and with that I was out of the office. I stopped by the director of maintenance office to explain what happened. I showed the director of maintenance the boiler efficiency charts and pointed to the graph that shows that the boiler will indeed reach 99% efficient if the return water is below 70 degrees F and the burner firing rate is at low fire. It is difficult to raise the temperature of a room using 70 degrees F water. The engineer used my competitors 99% efficient condensing boiler for the project.

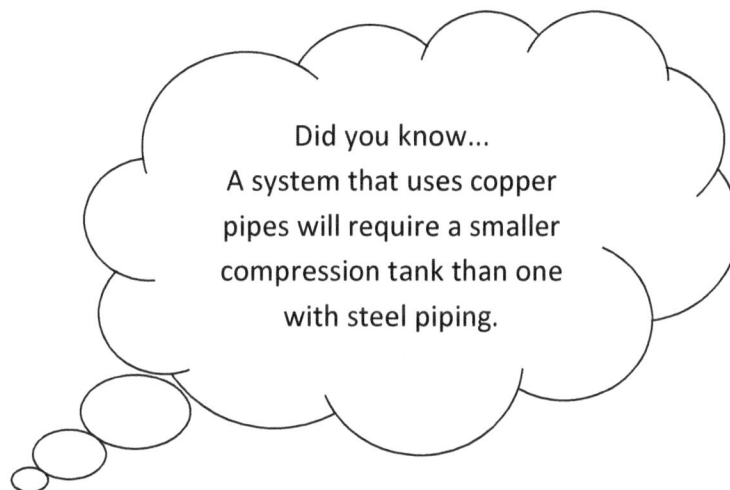

Did you know...
A system that uses copper pipes will require a smaller compression tank than one with steel piping.

5

Some things you need to know:

Whether you are designing or installing a boiler, there are certain things that you should know when replacing the heating plant. Consider it your due diligence.

Which areas are the hardest to heat? Always assume that there are indeed problems in the existing system because you will own any problems when the new boiler is installed. "We never had those problems before you replaced the boiler." is a common phrase. This will also separate you from the incompetent installer or designer. If it is a substantial repair, you may be able to add a disclaimer to your proposal absolving you of the responsibilities of the problem. We had a building where the original installer had installed the three way valves backwards. We got the job because we asked questions and provided a solution.

Why is the client replacing the boiler? Is it energy costs? If the client is using energy costs as the purchasing reason, they may be disappointed as the cost to replace a boiler easily dwarfs the savings in most cases. It may take twenty years to pay for the replacement boiler. This should be discussed up front.

System problems? If the distribution system has some issues, the replacement boiler may not resolve those problems. You will then have a client that spent lots of money with the unacceptable results. They will not be happy. We had a customer that wanted to replace a boiler because the top floors of the building were cold. His service company told him that the boiler was too small. We found that the system had insufficient water pressure and air bound radiators. We raised the system pressure and replaced the automatic air vents. The system worked great. The client decided that he did want to replace the boiler after our repair. That may seem to be a self sabotaging idea but the client turned out to be a rainmaker for our firm. He uses our company for service on all his boilers and had us replace the boilers in three other buildings. In addition, he tells all his friends to use us. The old boiler is still there but we have made a friend and an avid supporter of our company.

Reliability? If reliability is important to the customer, you might suggest modular or multiple boilers which offer some backup in the event of a malfunction.

Does the client have a boiler preference? If the client likes cast iron boilers and you sell copper boilers, this should be discussed prior to you investing much time on the proposal.

Can one boiler handle the load? If the existing heating system had two boilers, this is a very pertinent question. This will help you when sizing the new heating plant. Typically, old boiler rooms were sized the boilers for 66-75% of the load. In that way, if one boiler failed, the building would still have some heat, depending on the time of year. The onsite person will be able to give you some guidance on that.

Do you have a budget number for this project? You need to know how much the client would like to spend on their new heating system. My father used to say," You have steak tastes on a peanut butter budget." Are you thinking of a high end system complete with new pumps and compression tanks and your client barely has barely enough to replace one boiler? If the client balks at giving you a price, you could try bracketing the project. It may sound something like this, "A new heating system will cost anywhere between $50,000 and $200,000. Can you tell me if your budget is closer to the $50,000 or the $200,000?" This simple question will help you design a system that is within the budget parameters. What happens if the client balks about the lower price? Now is the time to work on that rather than after you spend all the hours to design and price a replacement system.

What is in the boiler room?

Ray's Rule #7 Always assume that the old boiler is installed incorrectly. I have used this credo throughout my career. Although the majority of the boilers have been installed properly, there have been many that were not. Those boilers that were not installed correctly would have been very costly to fix had we not seen the issues prior to installing the new ones. That leads me to the most important part of any boiler retrofit, the boiler room walk through. What will you connect the new boiler to? If you would like to avoid lawsuits, always assume that old boiler is incorrect.

Clearance The boiler room seems to be the repository of any junk, scrap, old papers, broken parts, and garbage. You have to decide where you will be installing the new boilers. Each boiler requires a certain amount of clearance around it for service. Is there junk around the boiler? What about papers stacked next to the boiler? A school in my area would store the paper student records next to the boilers. One day, the burner malfunctioned and the flame rolled out, igniting the boxes of student records. Check the installation manual and local codes for how much clearance is required around the boilers. It is typically about 12-30".

Exhaust Fan During the asbestos abatement project, all the insulation was removed from the steam piping. The boiler room was so warm that the classroom above the boiler room had melted

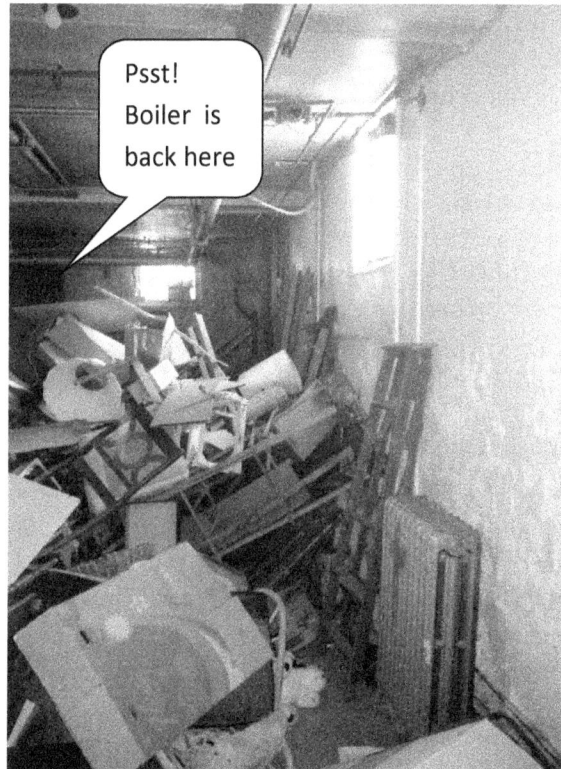

Psst!
Boiler is
back here

floor tiles. In an effort to cool the boiler room, someone installed a thermostatically operated exhaust fan. The original combustion air louvers were sized for the boiler only and were too small when the exhaust fan operated. The exhaust fan pulled the room into a negative condition and actually pulled the flue gases from the boiler. This filled the boiler room with a dangerous mixture of toxic gases. It also caused the boiler to soot, which exacerbated the issue. The carbon monoxide alarm sounded as the technician entered the room. He was able to push the "Stop Boiler" button at the entrance and shut down the boilers. The steam piping had to be reinsulated and exhaust fan disabled. The exhaust fan in the picture was not from the above project.

Boiler Room Heat Once upon a time, there was a quiet little kingdom that used boilers to heat the buildings. The boiler room where these modern marvels were situated was usually a toasty place. The king entered one of these boiler rooms to meet with the common folk, felt the heat and wailed that this was wasteful as he perspired for the first time in his life. He decreed that all boiler rooms should not be so warm so he tasked his staff with coming up with a solution. They suggested that they isolate the idle boilers, increase the insulation on the pipes, increase the combustion air opening size and change the flue from single wall to double walled vent pipe, much to the chagrin of the installers that complained about the extra costs. The law of the land was enacted and the citizens soon realized two unintended consequences. First, the boiler room temperature did drop, quite a bit. In fact, it was so cold in some boiler rooms that the gas valves would not open, making it even colder. The second consequence was that the boiler efficiency dropped by using the cold, denser air. So the king added an adendum to his decree that all boiler rooms should also have heaters in the boiler room to make it toasty once again. All was well in the kingdom. Long live the king.

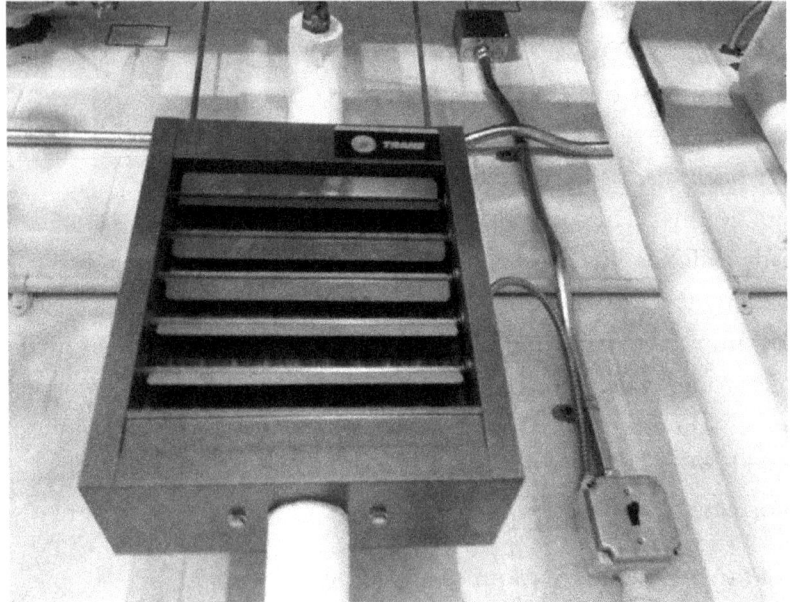

"Ask the custodian what he thinks about your new boilers?" the installer said with a grin. "Why?" I asked.
"You will see." he said with the same grin. I asked the custodian what he thought of the new boilers and he said, "I don't like them." I asked why and he just himmed and hawed and finally told me the real reason for not liking my new boilers. Everyday for the twenty years he worked in the boiler room, he would place the soup his wife made under the boiler and the radiant heat from the boiler would heat it. Now, there was no where he could heat his soup. I thanked him for his candor and purchased a small microwave oven for him. He liked my boilers after that.

8

Air Conditioner If the boiler room has an air conditioner, the International Mechanical code requires that either the combustion air has to be directly vented to each fuel burning appliance or the room has to have a refrigerant monitoring system that will detect and alarm if it senses leaking refrigerant. This could be expensive if you have to include it later.

Asbestos is a danger inside a boiler room as inhalation of asbestos causes several serious illnesses including lung cancer, mesothelioma and asbestosis. If the existing heating system was installed prior to 1972, it most likely contains asbestos. The asbestos could be on the flue or piping. It was typically used at the pipe fittings. Asbestos looks like a bright white flaky substance. Most facilities have had the insulation tested and should be able to inform you if any asbestos is in the boiler room. Some of the older cast iron boilers used asbestos rope between the sections or on the flue collector to seal in the boiler flue gases. In most instances, the company performing the asbestos testing in the facility would not know that and could have omitted it in their report. If you have any doubt, always have it checked. A safe idea is to add a disclaimer to your proposal that asbestos abatement is not covered in your initial quote. The remediation costs could be very high.

Dissimilar Metals When you mix black iron pipe with copper, it could cause electrolysis or galvanic metal corrosion. This corrosion generates a small electrical current as a result of the two dissimilar metals and an electrolyte, water. This electrolysis can cause erosion of the metal, typically the black iron fittings. If you must join the two metals, a dielectric fitting should be used. The dielectric fitting separates the two metals and eliminates the chance of the galvanic corrosion.

Invite the Boiler Rep When you are looking at a boiler project, invite the local boiler representative to accompany you. He or she may provide much more experience and that extra pair of eyes may spot something you missed. They like getting out of the office also.

Boiler Room Checklist The following checklist is the one I use when looking at a boiler replacement project. It helps me to remember to look at all the things I used to forget.

Boiler Room Replacement Checklist

Building			
Address			
Jobsite Contact		Date	
Boiler Manufacturer		Boiler Type	
Model		Serial	
Age		Quantity	
Input		Output	
Efficiency		Voltage/Phase	
Burner Manufacturer		Fuel Type	
Model		Serial	
Input		System Pressure	
Temperature Setting			
Combustion Air Size		Breeching Size	
Door Opening Size			
Gas Pressure		Gas Pipe Size	
Pump Manufacturer		Pump Model	
Pump GPM/Head		Pump Voltage/PH	
Expansion Size		Compression tank Quantity	
Supply Pipe Size		Return Pipe Size	
Notes:			

	Yes	No		Yes	No
Gas Train Vented to outside?			Draft Controls?		
Backflow Preventer?			AC in Boiler Room?		
Pipes Insulated?			Exhaust Fan in Boiler Room?		
Venting Issues?			Water Meter?		
Teflon on Gas Pipe?			Asbestos?		
Relief Valve Leaking?			Boiler Reset Control?		
Relief Valve Piping Correct?			3 Way Valve?		

Typical Hydronic Piping Based on 20 degree rise

Pipe Size	GPM	BTUH	Pipe Size	GPM	BTUH
2	50	450,000	6	850	8,500,000
2 1/2	80	850,000	8	1,800	18,000,000
3	140	1,300,000	10	3,200	32,000,000
4	300	3,000,000	12	5,000	50,000,000

Let's Design a Perfect Boiler Room

In my seminars, I ask the attendees what should be included in the perfect boiler room. The most common consensus is that our boiler room should consist of the following equipment:

- ✓ Two condensing boilers, each with a high turndown modulating burner.
- ✓ Each boiler will be sized for about 75% of the heating load to allow for backup in case of a malfunction of one the boilers in the middle of winter.
- ✓ The boiler should have automatic valves that will isolate the idle boiler to limit jacket and stack losses. Isolation of the idle boilers is also required by the International Energy Conservation Code.
- ✓ Combustion air should be vented from the outside.
- ✓ The boilers should be combination boilers that also heat the domestic water.
- ✓ Variable speed circulators should be included to help us meet the criteria of the International Energy Conservation Code.

The only caveat I explain to the seminar attendees is that the new boilers will be connected to a piping distribution system that was designed several decades ago.

When I was in high school, I worked for a foreign auto dealer that sold high end sports cars. Their top of the line model was an amazing vehicle that was very powerful. You could run the vehicle in first gear up to 50 miles per hour. The top speed of the vehicle was about 250 miles per hour. Each press of the gas pedal seemed to emit raw, exhilarating power. Although it was a remarkable piece of automotive engineering, it did have a sort of Achilles heel. After a few months, the engine would soot and start spewing black smoke. The dealer was overwhelmed by the warranty claims and irate clients. To help diagnose the problem, the factory sent in one of the top service technicians. My job was to pick him up from the airport and deliver him to the dealership. While sitting in traffic from the airport, I asked the technician what caused the problem with the engine sooting. He looked around and waved his arms in a circular motion and bellowed, "This. This is the problem." I told him that I did not understand. He said to me in broken English, "You need to drive our vehicles, hard! They are like a, how you say it, a race horse. They are not designed to sit in traffic, to take kids to futball practice or grocery shop. They will die a slow death if they do." That is like running a condensing boiler connected to an old high temperature loop.

Most condensing boilers were based upon European designs and many still use the actual European heat exchangers. What you should understand when using a condensing boiler is that the heating systems across the great pond are designed using much lower operating temperatures, about 37% lower. They are usually designed around the following temperature; 55^0C or about 131^0 F as the design temperature on low temperature systems. Since that is the hottest temperature used, the boiler water temperature is usually lower than that for the majority of the heating season. Over here, our old heating systems were designed using 180^0F as our design temperature. At that temperature, the condensing boiler is not condensing. It is like that high end sports car sitting in bumper to bumper traffic. They need to be run at low temperatures to get the most out of them, the colder the better. Although the boiler is an integral part of a heating system, it is not the only part. You need to consider what the boiler will be attached to.

The following are the European Design temperatures for hydronic heating system as per Mr. Harald Prell with Viessman

European Design Temperatures				
	Supply		Return	
Application	Celsius	Fahrenheit	Celsius	Fahrenheit
Before 1980	90	194	70	158
After 1980	75	167	65	149
Low Temp	70	158	55 or50	122-131
Condensing Boiler	55 or 60	131-140	45	113
Radiant Floor	45	113	35	95

Various Heating Emitter Design Temperatures				
	Supply		Return	
Location	⁰C	⁰F	⁰C	⁰F
Denmark	70	158	40	104
Finland	70	158	40	104
Germany	80	176	60	140
Korea	70	158	50	122
North America	82	180	70	158
Poland	85	185	71	160
Romania	95	203	75	167
United Kingdom	82	180	70	158

Courtesy: Mr. Robert Bean

Celsius to Fahrenheit			
Celsius	Fahrenheit	**Celsius**	Fahrenheit
100	212	**65**	149
95	203	**60**	140
90	194	**55**	131
85	185	**50**	122
80	176	**45**	113
75	167	**40**	104
70	158	**35**	95

Design the New Heating Plant

Now that you have your questions answered and your due diligence performed, you can start the design of the new boiler room

Sizing the Boilers

Hydronic System Sizing	Steam System Sizing
Use Building Heat Loss	**Use Connected Load**

When sizing replacement boilers, steam and hydronic systems are completely different. To properly size a new hydronic system, a heat loss of the building should be performed. The new heating plant should be sized to meet the building's heat requirements but not oversized as building inspectors are sometimes asking for a copy of the building heat loss with replacement boiler applications so they can verify that the new heating plant is not oversized. The sizing can be done manually but most people use software for the heat loss. The burner in the picture is from a system where the boiler is grossly oversized. At this building,

the system loafs along at low fire for most of the heating system. As a result, the burner head is destroyed after each heating season from the excessive heat. To combat this problem, the burner requires more air low fire to push the flame away from the burner head. While this saves the burner head, the burner efficiency is lower due to the excess air.

When designing steam systems, they should be sized according to the connected load. This is done by adding all the heat emitters. This may be baseboard radiation or radiators. You then have to add a

13

factor for all the pipes as they would consume heat if they were cold. This factor is between 33-50% of the heat emitters.

Heating systems were traditionally sized for two large boilers, each sized for 75% of the load. In that way, the building would have heat in the event of a single boiler failure. The drawback to this thinking is that the heating system is oversized by 50% on the coldest day. In addition, the other components are also 50% oversized including: Piping, pumps, compression tank, electrical, natural gas, combustion air, and flue. At warmer temperatures, the over sizing percentage is even greater. Please see the chart below.

2 Boilers sized @ 75% of heating load with 0 degrees F outdoor design temperature		
OA Temp	Building Heat Loss %	Over sizing %
0^{0}F Design Temperature	100%	50%
10^{0}F	90%	67%
15^{0}F	80%	88%
20^{0}F	70%	114%
25^{0}F	60%	150%
30^{0}F	50%	200%
35^{0}F	40%	275%
40^{0}F	30%	400%
45^{0}F	20%	650%

What kind of Boilers? There is much debate about which type of boiler should be used. Whether you choose copper, steel, cast, aluminum, or stainless steel, I have found that most boilers will provide longer life based upon three important criteria; proper installation, regular maintenance, and operating the boiler within its design parameters.

Condensing or Non Condensing When the price of gasoline spiked a few years ago, I traded in my large 8 cylinder truck for a new small 4 cylinder SUV. I thought I would save hundreds of dollars per month in fuel due to the newer more efficient vehicle. At that time, my territory included West Virginia, Pennsylvania, and Ohio. I was averaging between 35,000 to 40,000 miles per year in travel. On my first trip to West Virginia, I set the cruise control for 5 miles over the 70 miles per hour speed limit. Once I hit the mountainous part of the trip and was climbing one of the hills on Route 19, the vehicle seemed to slow down. It eventually just kicked out of cruise control. I veered into the right lane and stomped the accelerator. I felt like I was driving the Little Train that couldn't. The small 4 cylinder engine whined and protested. The automatic transmission down shifted but the best speed I could do was about 45 miles per hour up the steep hills. I was being passed by old ladies in big Buicks. It was embarrassing. The vehicle barely lasted out of warranty and when the engine died, I traded it in for a vehicle with a larger engine that could handle the hills of my territory. The mechanic said I beat it to death by driving it how I did. When you decide that you are going to replace the existing boiler with a condensing boiler without checking the existing system, it could be like that small SUV attempting to traverse the large hills. In most likelihood, the old heating system was designed for 180^0F at the design at the outdoor design temperature. Let me exemplify that. In my area, the outdoor design temperature is 7 degrees F which is the 2% temperature. What that means is that the heating system should be able to maintain 68^0F in the building at that outdoor design temperature of 7 degrees F. The 2% means that the average heating season will be at or below the outdoor design temperature 2% of the winter. The original heating design most likely used 180 degrees as the supply water temperature to the building at the outdoor design temperature of 7 degrees. If the existing piping and terminal equipment is the same as when the original system was designed, the system will still require 180 degrees at the outdoor design temperature.

Let us consider whether a condensing boiler is right for our application. If you are operating a condensing boiler at 180^0F, the boiler will have an efficiency range of about 83-85%. This is the same efficiency levels as some non-condensing boilers. The condensing boilers are usually about 20% more expensive, require more maintenance, and have a much lower life than standard boilers. The condensing boilers may not hit 90% efficient until the water reaches 100 degrees or lower and this may only happen about 20% of the winter.

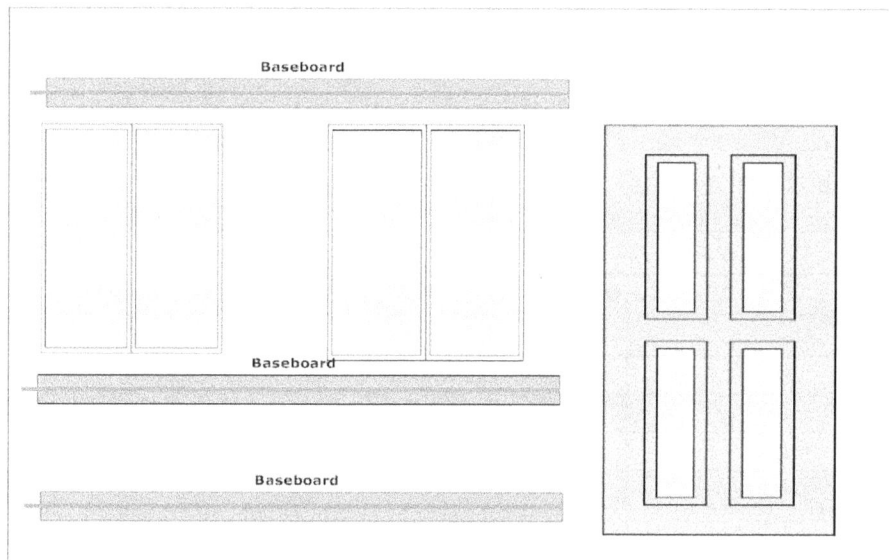

Check the heat emitters. Verify that the heat emitters, like the radiators or baseboard, are able to handle the load at the reduced temperatures. For the boiler to always condense, the radiators and baseboard will have to heat the space using 140^0F or lower water at the outdoor design temperature. If not, extra heat emitters may be required. This may be fiscally unrealistic as in some instances, you may need three times the amount of heating surface you currently have. You could have a wall filled with copper tubes. According to the design manual for Slant Fin model 351-10 copper baseboard, the baseboard will produce 870 Btuh using 1 gpm @ 180^0F. At 120^0F entering water, the same baseboard will only produce 310 Btuh, about 35%.

Ticking Me Off -When you require extra heating for a space, copper baseboard is often used. If improperly installed, it will make a ticking noise when the hot water enters the cool copper tube and many occupants find this annoying. Be sure to follow the manufacturer's recommendations for the proper installation. Some manufacturers use a plastic sleeve to limit the noise while others suggest that the pump operate continuously to avoid the ticking. If the latter is used, the operating costs will be higher for the building. Marshall Engineered Products in their book, Pocket Manual on Heating, recommends that straight baseboard radiation lengths over 30 feet long should be avoided unless provisions are made for expansion. The following are rules of thumb for baseboard heating lengths.

Baseboard Pipe Size	Recommended Element Total Linear Feet
1/2"	25 Feet
3/4"	70 Feet
1"	104 Feet

Why is my condensing boiler not condensing? Later in the book, I discuss how the fuel to air ratio affects the condensing temperature of the boiler. Let us look at a typical boiler where the manufacturer suggests a reading of between 8-10% CO_2 in the flue gas. That 8-10% reading correlates to between 3-7% oxygen. That means that with that CO_2 level in the flue, the boiler will start to condense somewhere between 133 - 123°F water, not 140. The dew point is the point at which the flue gases start to condense. The following is a chart to show you the condensing temperatures at different fuel to air ratios.

How Combustion Air Affects Boiler Condensing Temperatures			
O2%	CO2%	Excess Air %	Dew Point
3.0%	10.0%	15.0%	133
4.0%	9.50%	20%	131
5.0%	9.0%	29.0%	130
6.0%	8.40%	36.0%	128
7.0%	7.9%	46.5%	123

Is a Hybrid System Right For the Project? As heating costs are starting to rise, owners have to decide which boiler type is better for their building. The condensing boiler will reduce the fuel costs but has a shorter life than non condensing boilers. The non condensing boiler will last longer but the fuel consumption is greater. In addition, the maintenance costs are higher for condensing boilers than non condensing ones. What if you could combine the best of both? That is what a hybrid system does. We will combine condensing boilers with the non condensing. The standard efficiency boilers will be the lead boilers when the outside temperature is below 32°F. The condensing boilers will take over the lead when the outdoor temperature is above 32°F. In this way, you get longer life, lower maintenance costs, lower installed cost and increased system efficiency. In addition, the life expectancy of the condensing boiler is greatly extended because the run time is only half of the winter.

Hourly Occurrences If you are connecting your new boiler to a system that was designed for 180 degrees F, the condensing boiler may not operate in the condensing mode until the water temperature is below 140^{0}F. Assuming that we are using a typical One to One reset schedule, we may not hit 90% efficient until the outdoor temperature is above 50 degrees F. The following charts show the point at which the boilers are fully condensing.

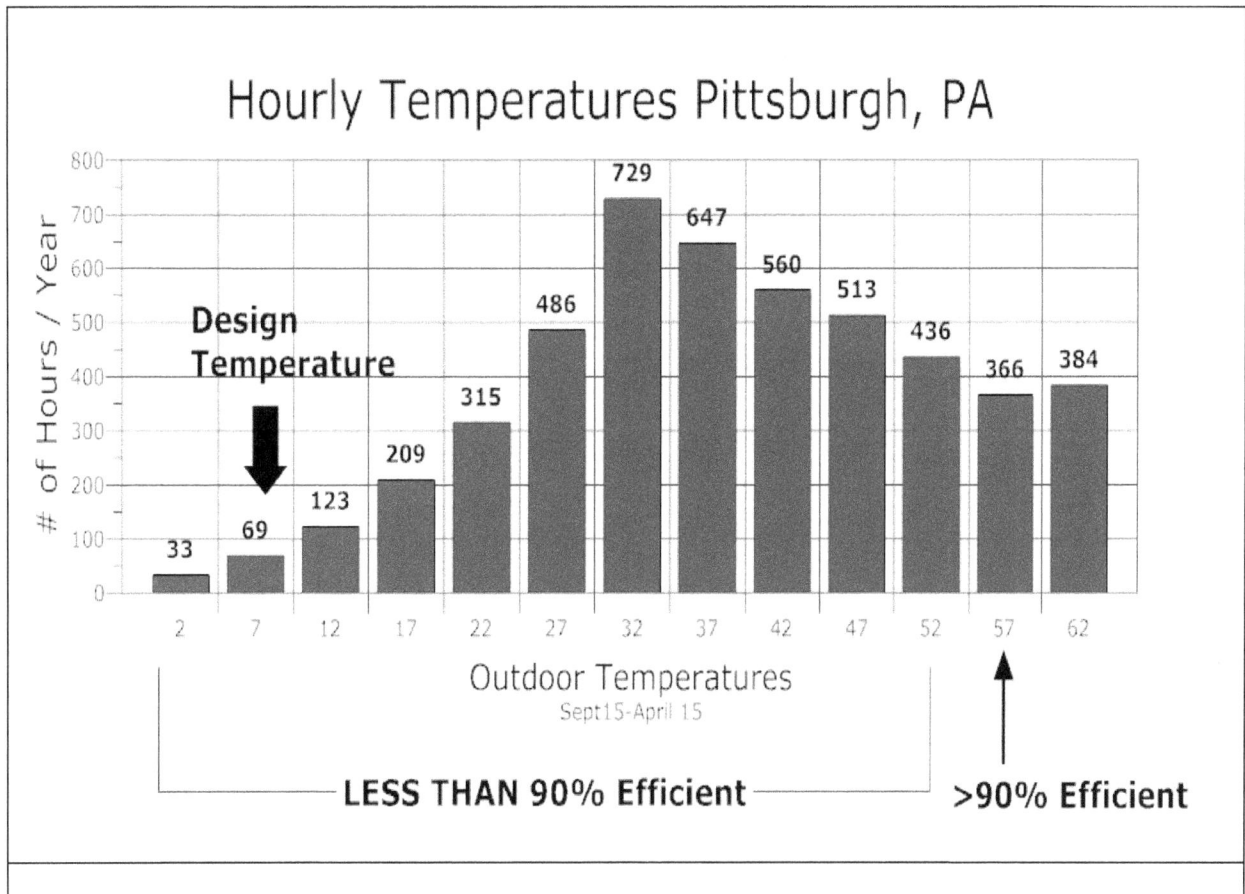

Hourly Temperatures Pittsburgh, PA

of Hours / Year

Design Temperature

Value
33
69
123
209
315
486
729
647
560
513
436
366
384

Outdoor Temperatures
Sept15-April 15

2   7   12   17   22   27   32   37   42   47   52   57   62

LESS THAN 90% Efficient ———— **>90% Efficient**

Hourly Temperatures New York, NY

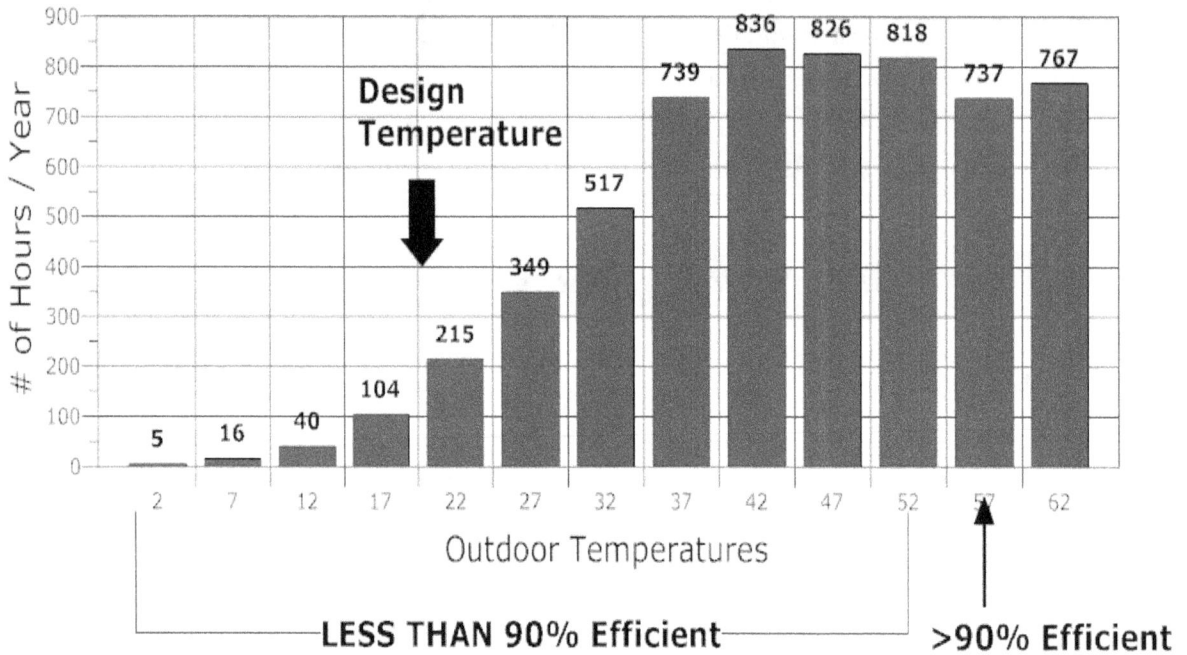

# of Hours / Year

Design Temperature

Outdoor Temperature	Hours
2	5
7	16
12	40
17	104
22	215
27	349
32	517
37	739
42	836
47	826
52	818
57	737
62	767

Outdoor Temperatures

LESS THAN 90% Efficient — **>90% Efficient**

Hourly Temperatures Charleston, WV

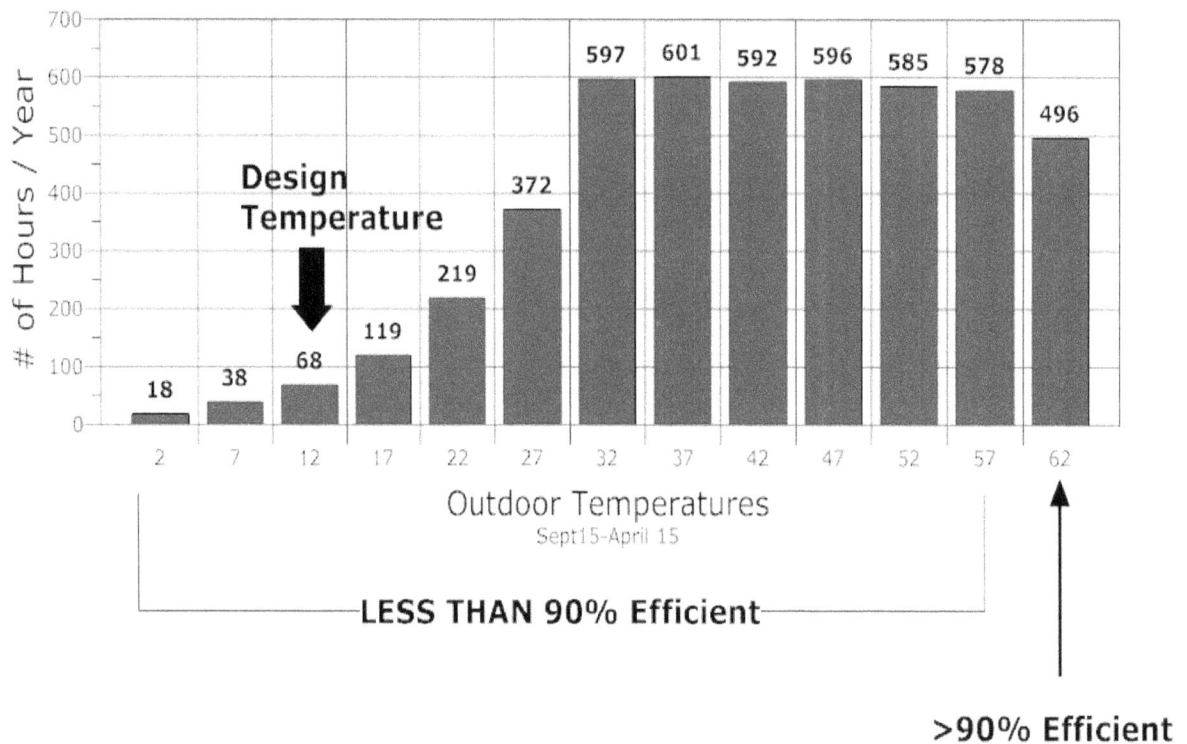

# of Hours / Year

Design Temperature

Outdoor Temperature	Hours
2	18
7	38
12	68
17	119
22	219
27	372
32	597
37	601
42	592
47	596
52	585
57	578
62	496

Outdoor Temperatures
Sept15-April 15

LESS THAN 90% Efficient

>90% Efficient

Sizing a boiler the old way. Boiler sizing was not as accurate as it is now. The following is how we used to size heating systems. Let us assume the building had a heat load of 1,000,000 Btuh. The designer did not want to get a call from the owner if the boiler failed in the winter so he or she sized two boilers for 75% of the heat load. This would be two boilers with an outlet of 750,000 Btuh each plus a sizing safety factor of 10% meaning that each boiler would be 825,000 Btuh output. The designer would find that the boiler manufacturer did not have a boiler that size so he opted for the next size boiler, which may be 900,000 Btuh each. The contractor gets a price from the wholesale house and the wholesaler tells the installer that they happen to have a great deal on two larger boilers because someone cancelled an order. The boilers are rated at 1,000,000 output but the contractor can have them for the same price as the smaller boilers. Since bigger was better, the designer agrees to the larger boilers. We now have a heating plant that is double the size than we need which is good because the owner will always have heat even if one boiler fails. To handle the bigger boilers, the pumps, piping, compression tank, combustion air, chimney, and wiring have to be larger. Now, our entire heating plant is grossly oversized.

Pipe Size Once you perform the new heat loss for the building with the hydronic boiler, verify whether the existing piping is large enough to handle the load. It is sometimes difficult to see the size of the existing pipes. The way I check the sizing is to look at the pipe fittings and or valves. The sizing is sometimes stamped on the side of the fitting. Flanges will have the sizing stamped on the side of the them. If all else fails, you could measure the actual size of the pipe. It is difficult to see the actual readings when holding a ruler or tape measure next to the pipe. In that case, you could wrap a piece of wire around the pipe and then measure the circumference. The chart below will help you when verifying the size of the pipes

Ray's Rule #8 Behind every successful construction project is a frustrated lawyer.

Pipe Dimensions			
Pipe Size	Inside Diameter Inches	Outside Diameter Inches	Circumference Inches
1/2	0.62	0.84	2 1/8
3/4	0.82	1.05	2 3/8
1	1.05	1.32	3 1/4
1 1/4	1.38	1.66	5 1/8
1 1/2	1.61	1.90	6
2	2.07	2.38	7 1/2
2 1/2	2.47	2.88	9
3	3.07	3.50	11
4	4.03	4.50	14 1/8
6	6.07	6.63	20 7/8
8	7.98	8.63	27 1/8
10	10.02	10.75	33 3/4
12	11.94	12.75	40
14	13.13	14.00	44
16	15.00	16.00	50 1/4
18	16.88	18.00	56 1/2
20	18.81	20.00	62 7/8
24	22.63	24.00	75 3/8

A condensing boiler with 4% oxygen in the flue will not start to condense until the water temperature reaches 131°F.

Water to Water? Getting back to my original project, the designer used a water to water heat exchanger to heat the domestic water and one to heat the swimming pool. While this is a good idea in areas where the water is hard, this does affect the efficiency of the system due to the heat transfer efficiency of heat exchanger. A water to water heat exchanger uses hot water inside the boiler to heat the ground water up to the desired discharge domestic water temperature. For commercial buildings with a kitchen, the discharge hot water temperature to the building should be

around 140°F. Many facilities are now using this temperature to kill any potential Legionella inside the pipes. To raise the incoming water temperature from the ground temperature to 140°F , the boiler water that is used to heat the domestic water has to be higher than 140°F. The boiler water temperature may be 20° F or more higher than the domestic water temperature. If that is the case, the boiler water temperature will be at 160°F or greater and your 98% efficient boiler will be about 84-86% efficient. When using a water to water heat exchanger, some municipalities require a double walled heat exchanger to assure that the boiler water will not mix with the potable water. Remember the sage advice from Dr. Egon Spengler in Ghostbusters, "Don't cross the streams."

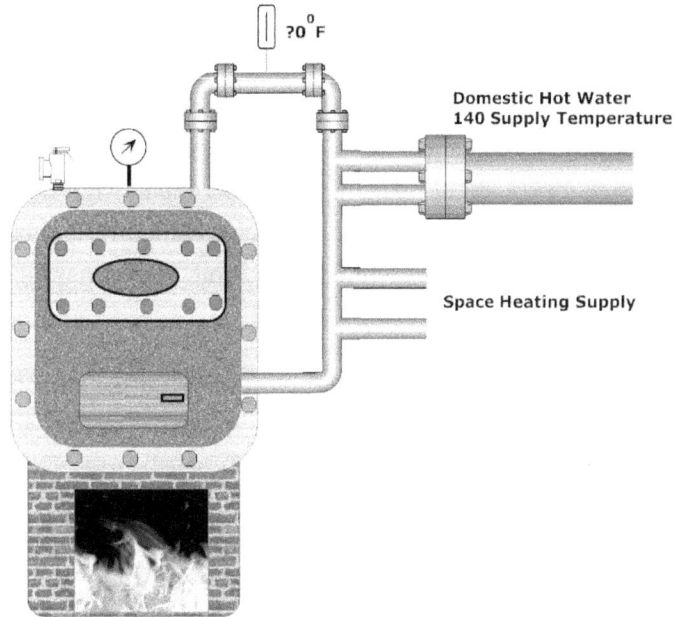

The pool heat exchanger will also require a warmer temperature than the discharge temperature to the pool. Assuming that we have a 110°F discharge temperature, our boiler water must be higher than that and it may be 10-20 degrees or more higher. The difference in efficiency is wide at that temperature. At 120°F degrees F, the condensing boiler efficiency will be around 90% and the efficiency drops to 86% at 130 degrees F.

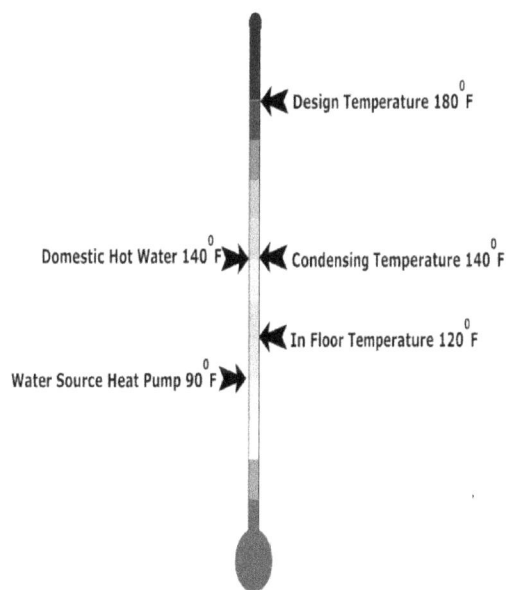

The thermometer picture shows the different temperatures commonly found in hydronic systems. The domestic water temperature is based on 140 degrees F that is common for commercial buildings with a kitchen.

Boiler Ratings The International Energy Code requires that you isolate the idle boiler and many designers will use isolation valves. This assures that the flow will not enter the idle boiler. There are a couple cautions that I should pass onto you if using isolation valves. The first is that the valves should have a time delay relay that will slow the closing for at least 10-15 minutes after the burner shuts off. If it closes too quickly, the boiler will continue to heat the water inside the boiler and could trip the manual reset temperature limit. If that happens, the boiler will not fire the next time there is a call for heat.

The second concern is that when the flow is shut off for one boiler, there could be excessive flow through the operating boiler. This may void the warranty and cause damage to the boiler. It could also affect the heat transfer ability of the boiler heating surfaces.

Another consideration is the sizing of the valves. They may restrict flow through the boiler, even when open. When sizing isolation valves, check the cV rating and pressure drop of the valve. The cV rating of the valve is the rated flow capacity of the valve.

The last thought is what happens to the pump when both boilers shut off. It will dead head the pump and could damage it. When the pump operates with the valves closed, the water could flash to steam. You may have to install a variable speed drive on the circulator or a pressure bypass pipe.

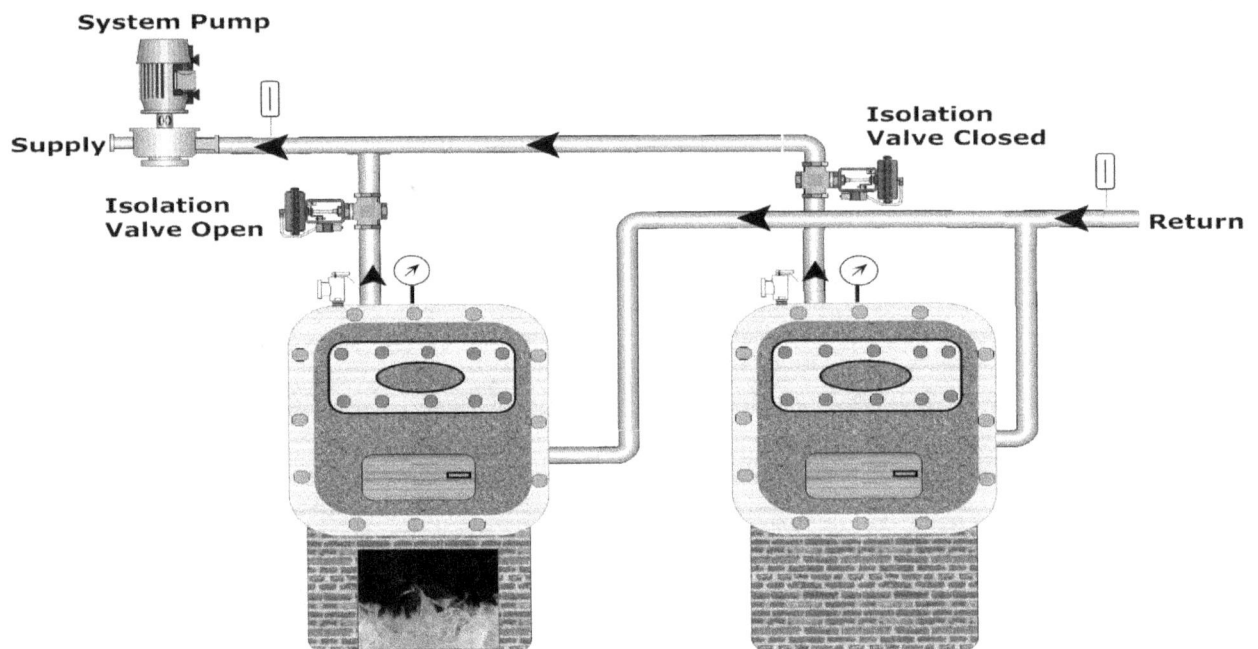

Do modulating burners really save money?

For years, we have designed boiler rooms consisting of two oversized boilers with fully modulating burners. The idea was that the boilers could "loaf" along at low fire because we were told that boilers were more efficient at low fire. I think it is time to rethink those strategies.

According to an EPA study entitled, Guidelines for Industrial Boiler Performance Improvement, "a boiler is most efficient when operating with a 50% to 80% load range." In addition, the study found that boiler efficiency drops dramatically if operated below the 50% load factor. In other words, operating a boiler at low fire is the least efficient way to operate the boiler. Another interesting part of the study was that operation of the burner with greater than a 2:1 turndown is not necessary for fuel-efficient operation. The study further suggested that a fully modulating

High Fire

Stack Loss 20%
200,000 Btuh

Actual Net to Bldg
770,000 Btuh
77% Efficient

Jacket Loss
3%
30,000 Btuh

1,000,000 Btuh Input

boiler operating continuously at 37% of the full load would be less efficient than a boiler operating on-off in the 50-80% firing range. I know that this may require a paradigm shift but bear with me while I explain.

Fully modulating burners are designed to safely operate throughout its firing range from high fire to low fire. The most common turndown ratings in commercial boilers range from 3-1 up to 10-1. Turndown is how far the burner firing rate can be lowered and still effectively fire. For instance, a 3-1 turndown burner means that the burner will be able to drop to 33% of its firing rate. A 10-1 turndown will be able to reduce its firing rate to 10%. High turndown is used to reduce the burner cycling and maintain a consistent temperature or pressure in the boiler. This is crucial if the boiler is used in an industrial process that requires a consistent temperature or pressure. It is not as crucial in a commercial space heating environment except to the control tech that can and will track the temperatures to within four decimal points of the set point.

When a power burner starts, it has a pre-purge period which ranges from 30 seconds to several minutes. During this pre-purge, the burner fan pushes air throughout the combustion chamber for a period of time required to provide four air changes inside the combustion chamber. This is to "purge" the combustion chamber and flue of any leftover combustibles. Some of the older boilers required seven air changes which prolonged the process. Many of the older, larger boilers also had a post purge which pushed air through the combustion chamber after the call for heat ended. While this air movement through the boiler does take heat from the boiler and send it outside, it is less wasteful than oversized boilers idling at low fire.

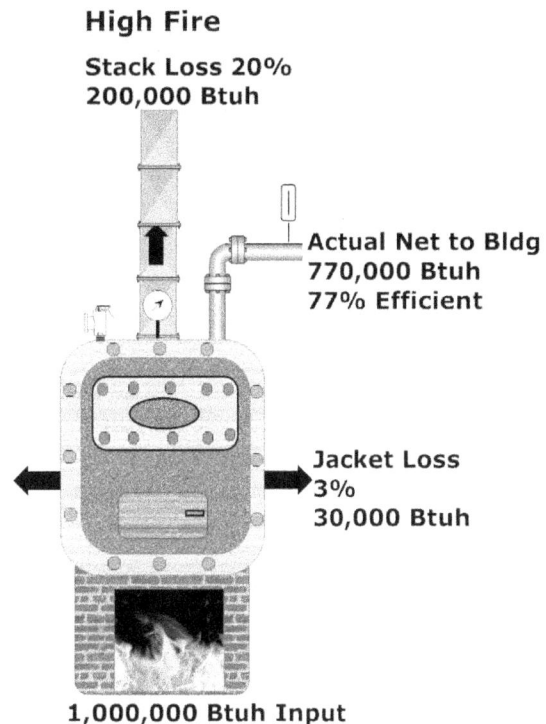

One of the other considerations is that with a high turndown burner, the blower motor will operate longer than a burner that heats quickly and shuts off, increasing electrical consumption.

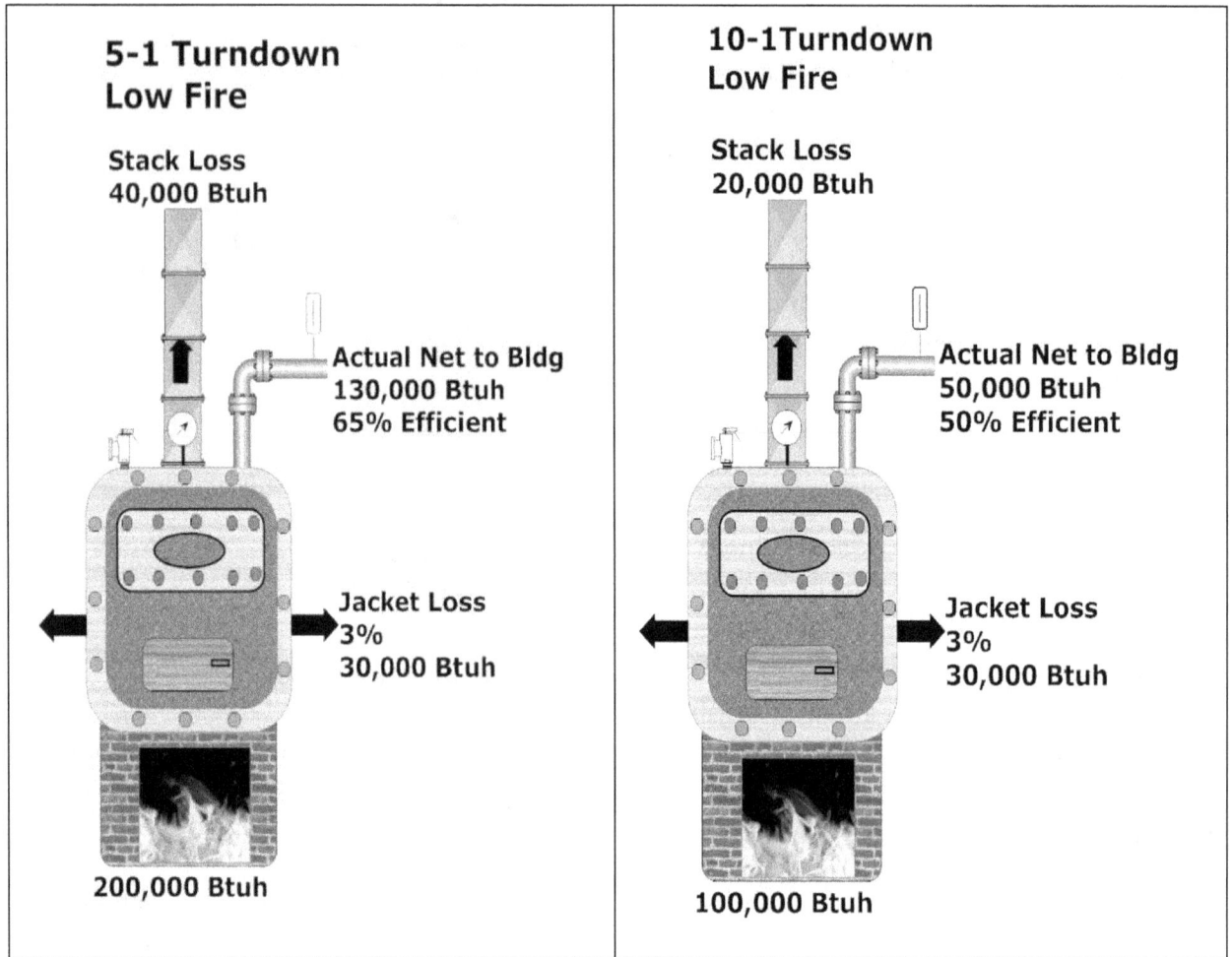

5-1 Turndown Low Fire

Stack Loss
40,000 Btuh

Actual Net to Bldg
130,000 Btuh
65% Efficient

Jacket Loss
3%
30,000 Btuh

200,000 Btuh

10-1Turndown Low Fire

Stack Loss
20,000 Btuh

Actual Net to Bldg
50,000 Btuh
50% Efficient

Jacket Loss
3%
30,000 Btuh

100,000 Btuh

Jacket Loss Traditional estimates suggest that a boiler will lose between 2-5% of its rating through the boiler wall into the boiler room at any time that the boiler is warm. Let us assume that we have a boiler with a rated input of 1,000,000 Btuh and it has a jacket loss of 3% from the boiler into the boiler room. A Btu is about the same amount of heat you would get from a wooden match stick burning. That equals 30,000 Btuh for every hour that the boiler is warm. Now, let us consider what happens when the burner drops to low fire. The boiler will still lose 30,000 Btuh because the boiler jacket does not

$70 \degree F$ $140 \degree F$ $70 \degree F$

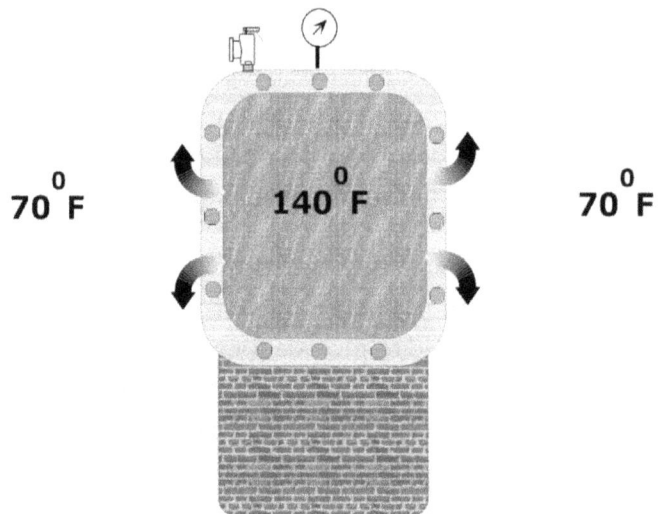

know that the boiler is at low fire. It just knows that the boiler is warm and the room is cooler. Nature loves equality. If our burner is a 3 to 1 turndown and drops to a low fire setting of 330,000 Btuh, that 30,000 Btuh loss just Tripled our jacket loss percentage to 9% of the firing rate, dropping the efficiency of our boiler from 80% down to 71%. If we have a 10-1 turndown burner, our jacket loss percentage just jumped to 30% of our firing rate. Our boiler efficiency is now at 56%.

If you have another boiler in the boiler room that is not isolated from the system, you will have jacket loss from that boiler as well because the boiler is warm as well. This could drop our efficiency even lower. For example, if our system had two boilers at 1,000,000 Btuh and only one was firing at low fire or 33%, our jacket loss would double to 60,000 Btuh and our system efficiency would be 62%. This is why the International Energy Conservation Code calls for isolation of the idle boilers.

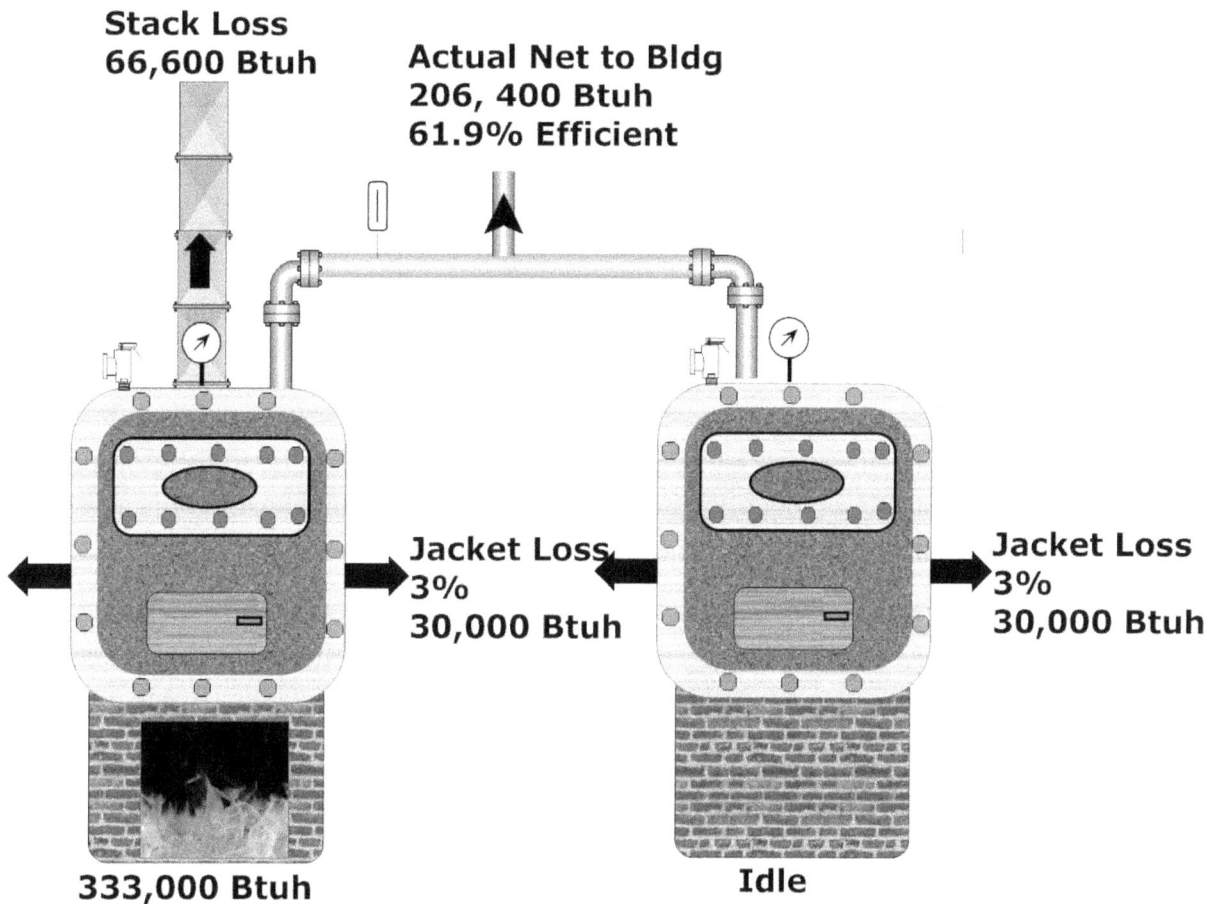

Stack Loss
66,600 Btuh

Actual Net to Bldg
206, 400 Btuh
61.9% Efficient

Jacket Loss
3%
30,000 Btuh

Jacket Loss
3%
30,000 Btuh

333,000 Btuh

Idle

The following will allow me to show you the heating plant efficiency with both standard and condensing boilers at low fire. We will compare two boilers each sized at 1,000,000 Btuh input at 80% and two at 90% efficient. Each boiler will have a 3% jacket loss into the boiler room.

Two 80% efficient boilers		
Burner Turndown	3-1	10-1
Input @ low fire	333,000 Btuh 33%	100,000 Btuh 10%
Flue Loss @ 80% efficiency	66,600 Btuh	20,000 Btuh
3% Jacket Loss from 2 boilers	60,000 Btuh	60,000 Btuh
Net to Building	206,400 Btuh	20,000 Btuh
System Efficiency	61.98%	20%

Two 90% efficient boilers		
Burner Turndown	3-1	10-1
Input @ low fire	333,000 Btuh 33%	100,000 Btuh 10%
Flue Loss @ 90% efficiency	33,300 Btuh	10,000 Btuh
3% Jacket Loss from 2 boilers	60,000 Btuh	60,000 Btuh
Net to Building	239,700 Btuh	30,000 Btuh
System Efficiency	71.98%	30%

According to Honeywell in their book entitled, Flame Safeguard Controls, a Low High Low burner is 10-15% more efficient than a modulating burner. A Low High Low burner uses a firing rate control like a modulating burner but only has two settings, Low or High fire. The burner will go between the two settings to meet the needs of the facility.

How many Boilers? I have found that several smaller boilers, piped primary secondary, with low high low burners can offer better seasonal efficiency than two large boilers with modulating burners in many instances. Each boiler room will be different so keep an open mind on your next replacement.

So how many boilers should we install? I believe that the sweet spot is between four to eight boilers. If we have four boilers each sized at 25%, the building would still have 75% backup in the event of a boiler failure just like the two boiler model above but the entire physical plant including the boilers, pumps, flue, combustion air, electrical and piping would be 50% smaller. If you are a little concerned about the sizing, you could have fours boilers sized at 30%. You would have 90% backup in the event of a failure and the heating plant would have a 20% over sizing factor rather than a 50%.

Courtesy of Triad Boiler Systems

If you consider eight boilers, the savings is even greater. Each boiler could be sized for 12 1/2%. If one failed, the building would have 87 1/2% backup. We sold eight of our boilers to a bank and the manager of the physical plant called one day and was concerned because only one of the eight boilers was firing on a very cold day. I asked if he had heat in the building and he informed me that he did. I assured him that it was operating within the design parameters. The boilers had heated the loop and one was maintaining it. The other boilers were isolated. He was impressed.

Front End Loading Should we replace all of the boilers? Earlier, I suggested that you ask the client how much money they have for the project. If money is really tight, you may consider a technique called Front End Loading. No, this is not something female movie stars do with plastic surgery. It is where you leave one of the old boilers in place. The boiler you are removing can be used for parts for the remaining old boiler. You then install one to several boilers that are to handle the brunt of the load and the old one is there as a backup in the event of a failure or severe weather conditions. This is sometimes referred to as a "Summer Boiler" as this boiler is used for heating the reheat loops in the Summer.

Durability How long will the new boiler last? That is a question I am often asked by the building owner or engineer. The answer is that it is difficult to ascertain. It was not uncommon to see boilers that are over forty years old still working in a facility. I am not sure how long the new boilers will last. According to ASHRAE, steel and cast iron boilers will last between 24-35 years. Because they are relatively new, condensing boilers do not have a projected life here in the United States. I

27

consulted the Chartered Institute of Building Service Engineers or CIBSE in the UK and they suggest that condensing boilers have an estimated life of 10-15 years. As with anything, maintenance plays a great part in the longevity of a boiler. This would include water treatment. ASHRAE also estimates the life expectancy of the following heating components:

Source ASHRAE

Equipment	Years
Boilers, Standard	24-35
Burners	21
Boilers, Condensing	10-15*
Pumps, Base Mounted	20
Pumps Pipe Mounted	10
* Based upon findings by Chartered Institute of Building Service Engineers	

Maintenance Costs According to ASHRAE, the steel fire tube boiler has the lowest maintenance costs of the traditional style of boilers, followed by a water tube boiler. The cast iron boiler has the highest maintenance costs. They did not publish the maintenance costs of condensing boilers but I believe they would be higher than the above three. ASHRAE has a really cool website where you can see the longevity and maintenance costs of any type of HVAC system. The information is supplied by the ASHRAE members. It is called ASHRAE: Service Life and Maintenance Cost Database at http://xp20.ashrae.org/publicdatabase/

Low mass boilers Many of the newer condensing boilers are what is referred to as low mass boilers which means that the water volume is low. Light load conditions sometimes cause short cycling and sooting of the fireside of the boiler. To eliminate that possibility, many installers install a buffer tank that will allow longer run times and higher efficiency. The buffer tank is simply a storage tank that adds volume to the heating loop which minimizes short cycling of the boiler. Another consideration with low mass boilers is that the boiler room will require heat in cold climates. I heard about a housing authority in Chicago, IL that had their low mass boilers actually freeze during one particularly cold winter. The cold air came in the flue.

Table Tennis versus Hydronics When I was an apprentice, I used to work with a skilled service technician that told me a story I never forgot. He was a laid back, quiet guy. He was called to a jobsite for an intermittent problem. Every day, a different room would not have heat and he was really at the end of his patience as was the owner of the building. It was beyond his expertise as everything worked the way it was supposed to. Whether it was fate or his patience, he was there when the system malfunctioned. On a whim, he unscrewed the union and opened the pipe to see what was blocking the flow to the room. Inside, he found a dirty ping pong ball. This thing would go along with the water flow and toward the next open valve. At night, when the pump shut off, it would be poised waiting for its next victim, an open valve. The ping pong ball found its way into the piping because one of the installers was mad at the boss and thought this would be the perfect revenge.

28

Connecting to the Piping

Try to imagine what life was like in the late 1800's in the United States, specifically 1899. In that year, voting machines were first approved by congress for federal elections. The first lawnmower was patented. Queens and Staten Island merged with New York City. Willis Carrier would not invent air conditioning for another three years. It was a very significant year in heating system history. At that time in history, boiler explosions unfortunately were a common occurrence. During that year, a collection of boiler manufacturers that was formed to provide "cooperative competition" in the industry, entitled the Carbon Club met at the Murry Hotel in New York City. On their agenda was a vote on two standards that are still with us to this day. The first item was that all comfort low pressure steam heating systems should be designed using 2 pounds of steam pressure as the standard operating pressure for comfort heating boilers. That means that buildings with low pressure steam systems should be able to heat the building at the outdoor design temperature using 2 pounds of steam pressure or less, regardless of the building size. Did you know that the Empire State Building in New York City, at 1,250 feet high, heats the building using only 2 pounds of steam pressure? I challenge attendees in my classes that if a building that size could heat it using 2 pounds of pressure, they should be able to heat their building with the same pressure. When you see a building using steam pressures higher than that, realize that there are most likely problems in the piping system and they will be your problems if you design or install a new boiler. Prior to that standard, it was not uncommon to see steam pressures in excess of 60 psi used for steam space heating. The 2 psi standard made it safer for those with steam heat.

The second major standard they agreed to was to use $180^{0}F$ as the design water temperature for hydronic heating systems. From that date on, all buildings using hydronic heating systems were to be designed to be heated at the outdoor design temperature of their locale using $180^{0}F$ water. All the components and heat emitters were designed using that same temperature.

What does this have to do with your replacement heating system design? If you are replacing a furnace in a home

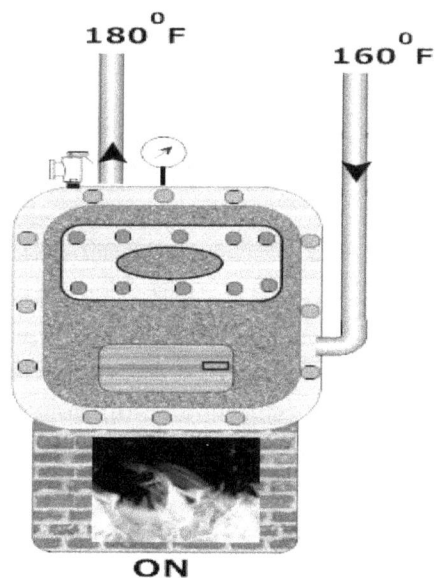

with a new one, a condensing furnace will make fiscal sense as that unit will condense any time it is operating due to the temperature difference between the return air from the room and flue gas temperature. When you connect a condensing boiler to an existing hydronic system, it may or may not condense. If the water temperature is above 140°F, the condensing boiler will not condense at that temperature. If the water temperature is at the 180°F design temperature, the condensing boiler will have efficiencies around 83-88%. In other words, the greater the heating load in the building, the lower the efficiency of the boiler. If the water is below 140°F, the condensing boiler will start to condense. Many people think that the boiler magically becomes 95% efficient as soon as the water hits that temperature. In reality, it may not even reach 90% efficiency until the water is below 100°F. I smile when I see these boiler ads showing banners touting, "Up to 98% efficient." If you read the fine print, the boiler hits that fabled 98% efficiency with 60-80 degree F water. In my area, 60 degree water in an old cast iron radiator will not provide much heat to the room. Since most of the existing boiler rooms in this country were designed using 180°F supply water standard at the outdoor design temperature, they may only condense for a fraction of the heating season. If the designer's calculations were correct, the heating system should be able to maintain 68 degrees F inside the building at the outdoor design temperature using 180 degree F water. *Please see the impact of water temperature on boiler efficiency.*

Impact of Water Temperature on Efficiency

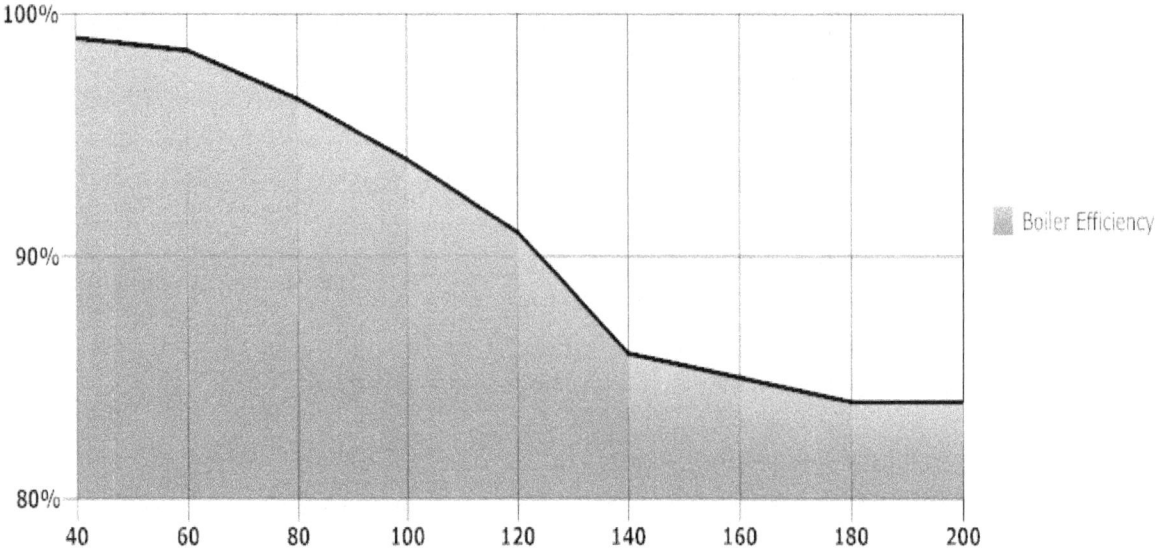

Delta T or temperature difference is also a crucial number for boilers as well. Most hydronic systems were designed using a 20 degree F delta T through the boiler. For instance, a boiler may be designed to heat water going through the boiler by 20 degrees F. A rise of greater than the design temperature rise could lead to damage of the boiler, or a phenomenon called "Thermal Shock." This thermal shock literally shakes the boiler to pieces due to rapid expansion and contraction. Try to imagine what would happen to a paperclip if you kept bending it back and forth. It would eventually fail. So does the boiler.

Terminal equipment such as fan coil units and baseboard radiation are also most likely sized for 180 degree design temperature with a 20 degree Delta T. The outlet will be cooler than the supply. The difference between the supply and return temperatures is the heat given up to the space through the coil.

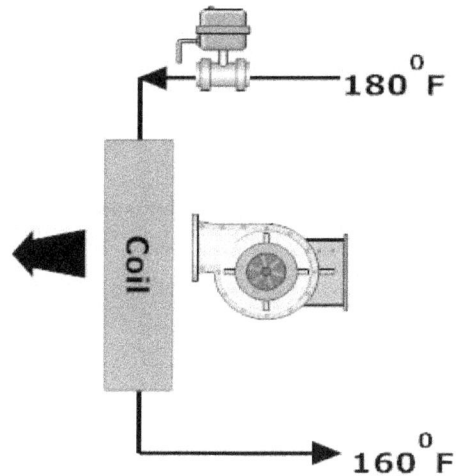

System velocity or the speed that water flows though the system is very important. If the flow is excessive, the system will be noisy and could actually erode the interior of the piping. If the flow is too slow, the air pockets will not be entrained in the water and could lead to blockages from the air, resulting in comfort complaints. This can occur with flows below 2 FPM. The flow should be 2-4.5 FPM or Feet per Minute in occupied areas and up to 6 FPM in unoccupied areas. Marshall Engineered Products, in their book Pocket manual on Heating, contend that velocities greater than 4 feet per second will cause noisy operation. If you are a first born and feel the need to calculate the velocity yourself, this is the formula:

$$\text{Velocity} = \frac{4 * flow\ rate\ (GPM)}{\pi * (pipe\ diameter)^2}$$

To further punish yourself, I have included the external and internal pipe diameters of common pipes. Use the internal pipe diameter to estimate the velocity. This is based on Schedule 40 pipe. In the rear of this book, I have the GPM for different velocities calculated.

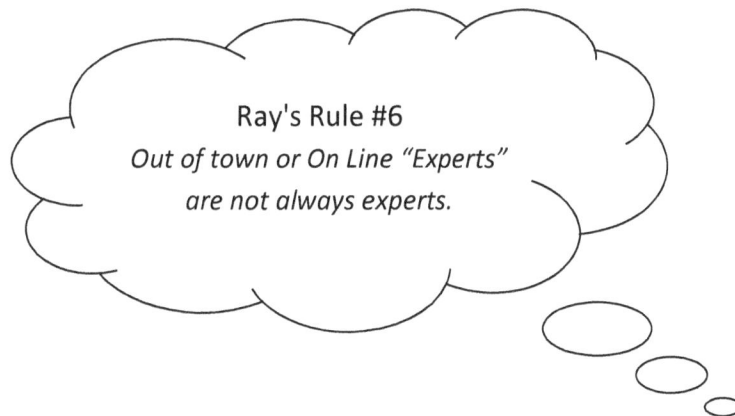

Ray's Rule #6
Out of town or On Line "Experts"
are not always experts.

Pipe Sizes

Pipe Size Inches	External Diameter Inches	Internal Diameter Inches	Pipe Size Inches	External Diameter Inches	Internal Diameter Inches
3/8"	0.68	0.49	6	6.63	6.07
1/2"	0.84	0.62	8	8.63	7.98
3/4"	1.05	0.82	10	10.75	10.02
1"	1.32	1.05	12	12.75	11.94
1 1/4	1.66	1.38	14	14.00	13.13
1 1/2	1.90	1.61	16	16.00	15.00
2	2.38	2.07	18	18.00	16.88
3	3.50	3.07	20	20.00	18.81
4	4.50	4.03	24	24.00	22.63

We received a frantic during a recent below normal cold snap from a nursing home maintenance supervisor telling me that our four year old boilers were not working. They were going to have to evacuate the seniors from the building in the frigid weather. While talking with him, I asked him what the temperature read on the boilers. He said that they were at 200 degrees F. He had turned them up from 180 degrees to get more heat. I kindly explained that if the boilers were running with 200 degree water, the boilers were working. He simply said, Help! I sent my technician to the building and he did verify that the boilers were supplying 200 degrees F supply water to the building. He further informed me that the temperature drop through the coils in the fans was only about 2 degrees F. He asked what to do. I was stumped, shrugged my shoulders, and suggested that he close the balance valves until he got a 20 degree drop through the coil. I heard the client yelling in the background. "He wants you to do what?" I have found that if you say something with authority, people will listen to you, even though I had no clue what to do. I asked the tech to try it on one unit. He called back excitedly in a couple minutes and said "It's working." As soon as he slowed the water, the discharge air temperature began to rise. He adjusted all the coils and the owner was ecstatic as they did not have to evacuate the building. The high velocity water would not give up its heat to the coil.

How many boiler rooms? I visited a building where they wanted to replace a boiler. As I did my walk through and calculations, the existing boiler would never be large enough to heat the building. I was befuddled and the jobsite contact told me that the boiler did heat the building. After scratching my head, I asked if there was another equipment room. Sure enough, the building had a second boiler room in the other wing of the building. Aw, the old hidden boiler room ploy.

The following chart is used for sizing a hydronic system. It includes suggested flow rates for different pipe sizes:

PEX Piping
Maximum Hydronic Flow Rates

Pipe Size	Maximum Flow GPM	Btuh
3/8"	1.2	12,000
1/2"	2	20,000
5/8"	4	40,000
3/4"	6	60,000
1"	9.5	95,000

Steel Pipe Based on 20 degree F Delta T

Pipe Size	Maximum GPM	Btuh
1/2"	2	15,000
3/4"	4	40,000
1"	8	80,000
1 1/4"	16	160,000
1 1/2"	25	250,000
2"	45	450,000
2 1/2"	80	800,000
3"	140	1,400,000
4"	300	3,000,000
6"	850	8,500,000
8"	1,800	18,000,000
10"	3,200	32,000,000
12"	5,000	50,000,000

Copper Pipe Based on 20 degree F Delta T

Pipe Size	Maximum GPM	Btuh
1/2"	1 1/2	15,000
3/4"	4	40,000
1"	8	80,000
1 1/4"	14	140,000
1 1/2"	22	220,000
2"	45	450,000
2 1/2"	85	850,000
3"	130	1,300,000

Piping Design There are a couple of basic designs that were commonly installed in hydronic systems. The first is a **Two Pipe Direct Return** system. In this system, the first radiator fed is the first returned. The drawback to this type of system is that it could short cycle. The water flow will follow the path of least resistance. Most of the flow and heat will go to the closest radiators. To eliminate the short cycling, the system may require balancing valves that regulate the flow so that all the radiators have enough heat, regardless of their distance from the boiler. The installed cost for this system is the lowest. If the radiators closest to the boiler are warm and the ones at the far end are cool, check the balance valves. You may have to close them slightly to force more water to the far end of the building. I have found that balance valves work best on the outlet of the heat emitters as it will get less air into the coil.

Two Pipe Reverse Return In this system, the installer adds an additional pipe so that the first radiator fed is the last returned to the boiler. This makes the system almost self balancing. The extra pipe increases the installation cost.

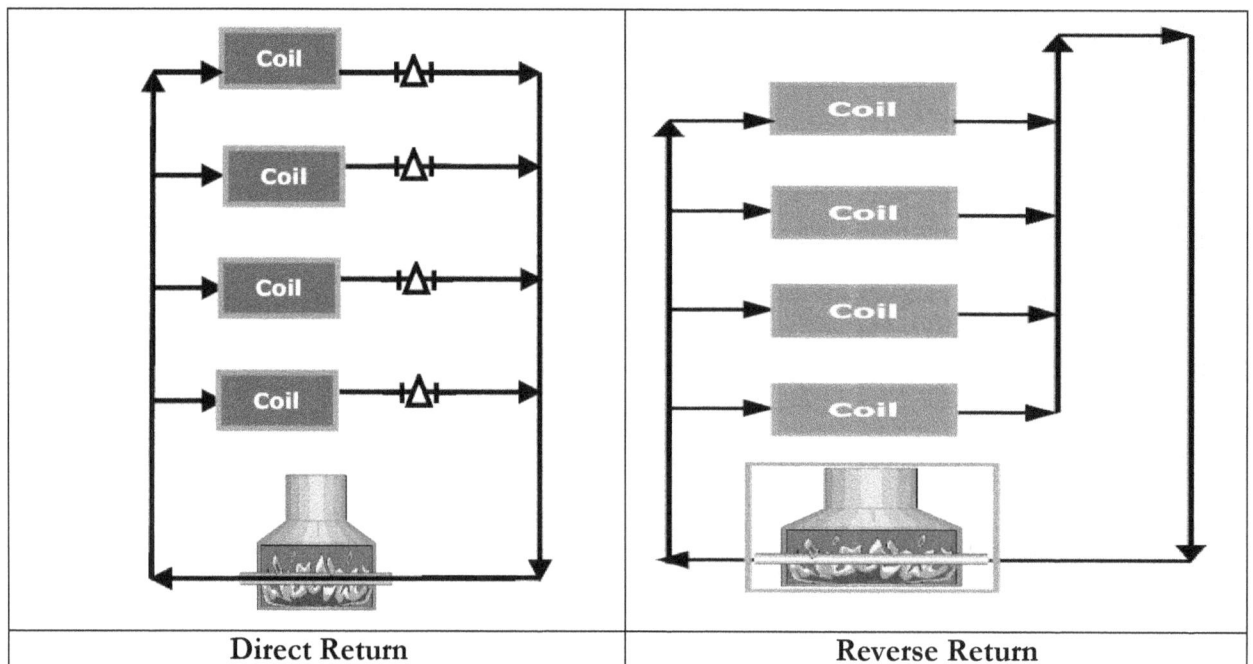

Direct Return	Reverse Return

Home Run With the advent of Pex tubing, we can have systems that allow home runs which make control and comfort easy. In this type of system, each terminal unit or zone can have its own supply and return. It makes zoning a snap, resulting in better comfort for the client.

Two Pipe Heating and Cooling A common complaint that building maintenance departments hear is that the occupants are either too hot or too cool. In some instances, both complaints may come from the same room. One of the systems with the most comfort complaints is a Two Pipe System. As the name implies, the system uses two pipes to the building; a supply and a return. In the heating season, the water in the pipes is heated with a boiler and in the cooling season, it is cooled with a chiller. During the peak of each season, this strategy works great. The comfort complaints come during the shoulder times of the year, like Fall or Spring. During that shoulder season, the building may require heat in the morning and cooling in the afternoon. The complaints occur due to the dead band of the temperature difference between the lowest hot water and highest chilled water temperature. Standard efficiency boilers are designed to operate above 140° F. Operation below that temperature could cause the flue gases to condense which can destroy the boiler and chimney. Conversely, most chillers cannot operate with temperatures above 90°F *(Please check with chiller manufacturer as to their temperature limits)* . So, we have a temperature difference of 50° F between the heating and cooling set point. You have heard that saying about how long it takes for a watched pot to boil, that is nothing compared to the time it takes for the temperature inside a two pipe system to drop from the minimum heating temperature to the maximum cooling temperature on a mild Spring or Fall day. It may be anywhere from several hours to a full day and by that time, we may need heat again. It causes many comfort complaints.

There are a couple ways to provide better comfort for your customer. The first is to use condensing boilers that can be operated at the lower temperatures. Some may be able to be operated down to the 90^0F temperature that the chiller can handle.

On a local project I had, the customer did not have the funds to install condensing boilers so we provided a combination boiler that had an internal copper coil. Normally, this internal coil was used to heat the domestic hot water. We used this as the first stage heat and it allowed us to have a heating loop temperature of 90^0F without flue gas condensation since the internal boiler water temperatures were above the condensing temperatures. The customer was able to switch from

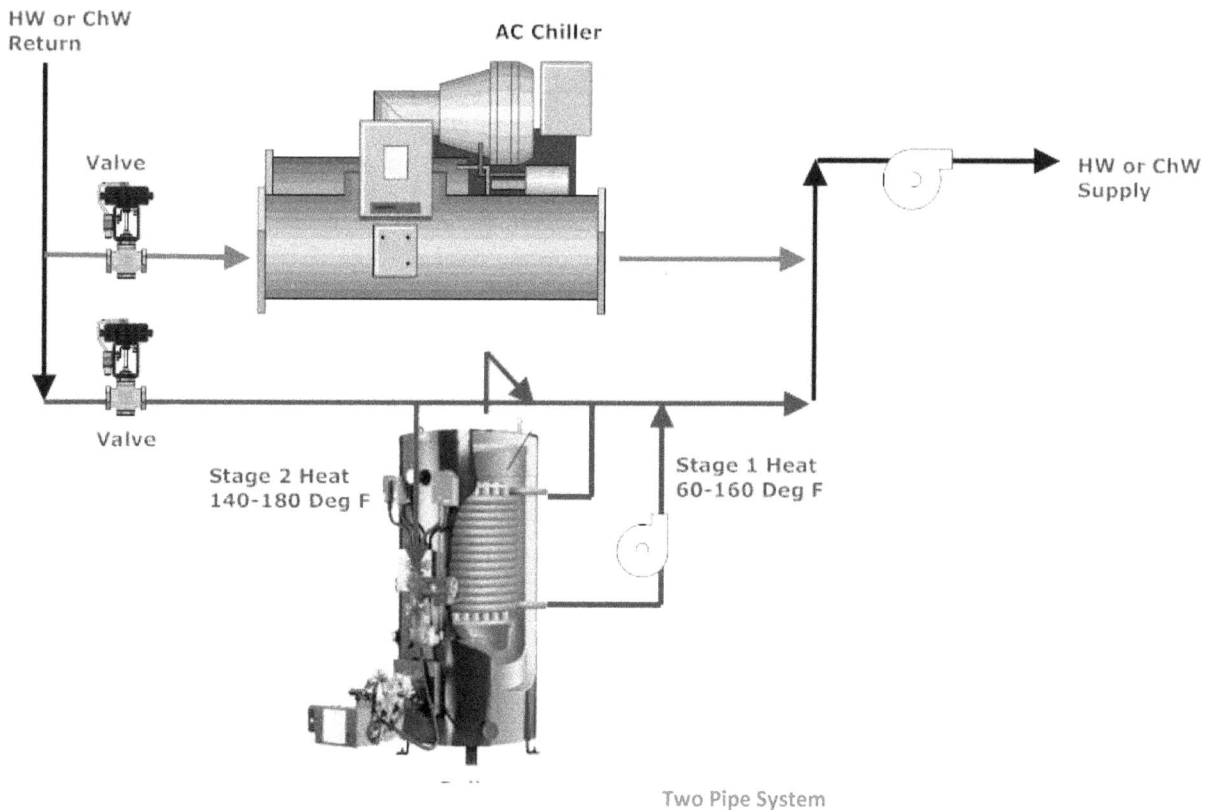

HW or ChW
Return

AC Chiller

Valve

HW or ChW
Supply

Valve

Stage 2 Heat
140-180 Deg F

Stage 1 Heat
60-160 Deg F

Two Pipe System

35

heating to cooling very quickly and reduced their comfort complaints.

If your state adheres to the International Energy Conservation Code, there are a couple of regulations that they require when controlling a two pipe system. According to the 2009 International Energy Code, the following must be done:

Section 503.4.3.2 Two-pipe changeover system.

"...systems that use a common distribution system to supply both heated and chilled water shall be designed to allow a dead band between changeover from one mode to the other of at least 15° F(8.3° C) outside air temperatures." A common set point arrangement is for the heat to be enabled below 50° and air conditioning enabled above 65° F outside temperature. In between those two set points, most buildings use a combination of economizer operation for cooling and return air for heating.

"...provided with controls that will allow operation in one mode for at least 4 hours before changing over to the other mode." This may require some planning for the building owner to avoid complaints.

"...provided with controls that allow heating and cooling supply temperatures at the changeover point to be no more than 30°F (16.7°C) apart." In other words, the heating loop shall be no warmer than 120° if the chiller is designed for 90° F entering water temperature at the changeover point. This is below the typical lowest operating temperature of most standard efficiency boilers.

Three Pipe In this type of system, there are two supplies and one common return. One supply is for the heating and one is for the cooling. They then feed into a common supply pipe to the coil and a common return from the heating and cooling units.

Four Pipe Each system has its own supply and return. There are four pipes run through the building which increases the installation costs. The terminal units choose whether to use heating or cooling. The advantage to this system is that it allows heating and cooling to be available at all times. The drawback is that you are operating heating and cooling at the same time. It is also the most expensive system to install due to the extra pipes.

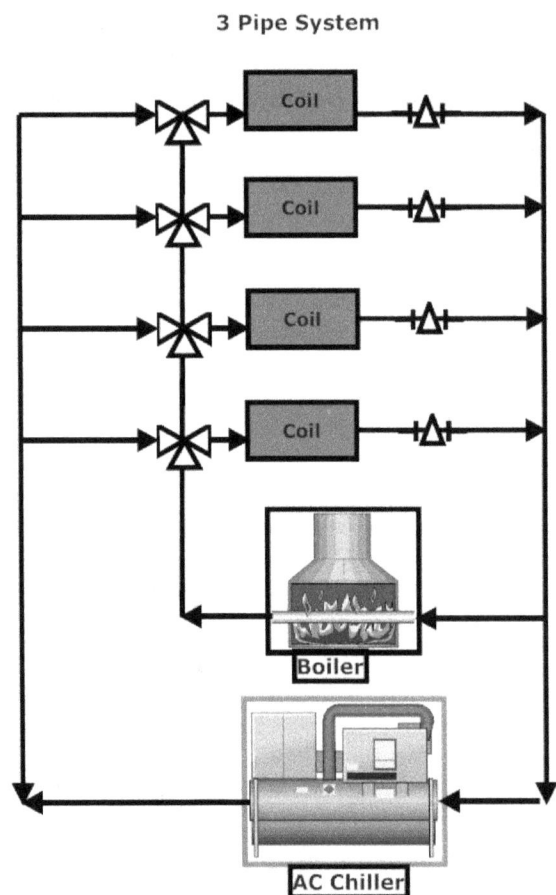

3 Pipe System

System Efficiencies Traditional old systems were piped to have flow through all the boilers all the time, even the idle ones. This led to higher jacket and stack losses. Jacket loss is the transfer of heat from the boiler water through the walls and jacket of the boiler to the mechanical room. There are a couple things that affect transfer of heat; the temperature of the water or steam inside the boiler, the ambient temperature of the equipment room and transfer medium such as boiler jacket thickness and insulation. Nature likes equality. It will try to equalize the temperatures. If a boiler has 180 degree inside and the boiler room is at 60 degrees F, the boiler will lose a certain amount of heat from the boiler into the boiler room. The only thing that changes the flow rate of Btus from the boiler to the boiler room is the temperature difference between the two objects and the material that separates the two. If the boiler has a better insulation, the temperature transfer is lowered. Lower temperature boiler water will decrease the transfer of the Btus as well. The funny thing about jacket loss that we seldom consider is that the heat will transfer anytime there is a temperature difference and the boiler jacket does not care whether the boiler is firing at high or low fire or even if the boiler is actually operating. If two boilers are in the same boiler room and are piped in the traditional flow through design and only one is operating, they will both have the essentially the same jacket loss as the water temperature will be the same in the boiler that is firing and the one that is off. There is disagreement in the industry as to heat lost through the boiler jacket into the boiler room was actually lost since it was still in the building.

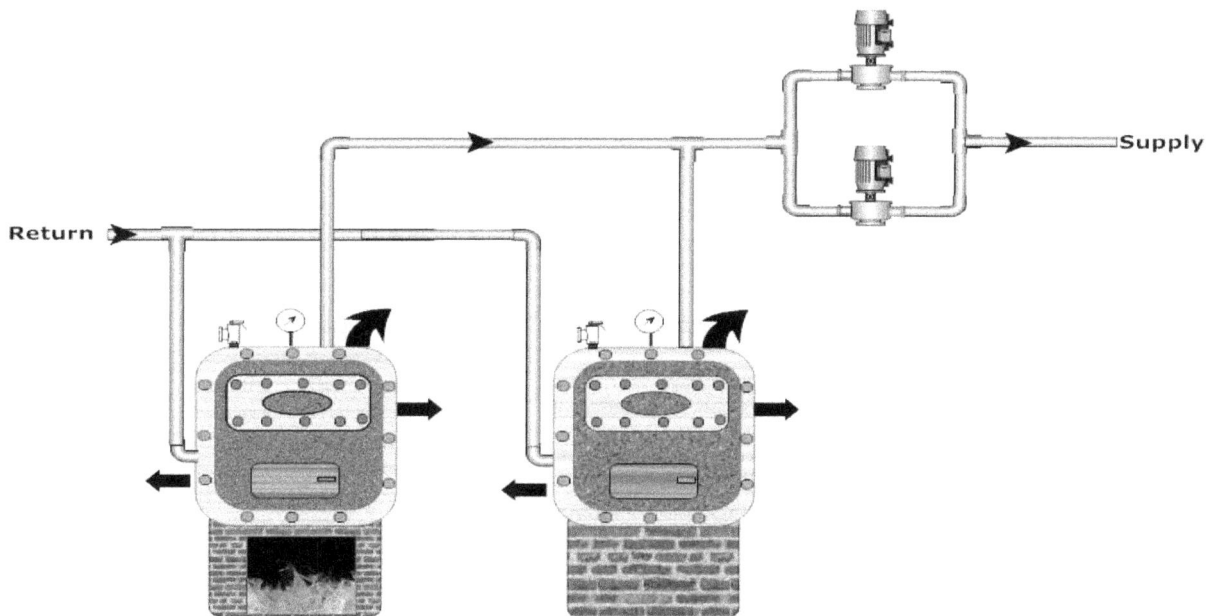

Stack loss This is the loss that transfers heat from the boiler to the cooler stack or chimney. To reduce this loss, many of the furnaces and boilers used stack dampers. These dampers open when there is a call for heat and close after the burner shut off.

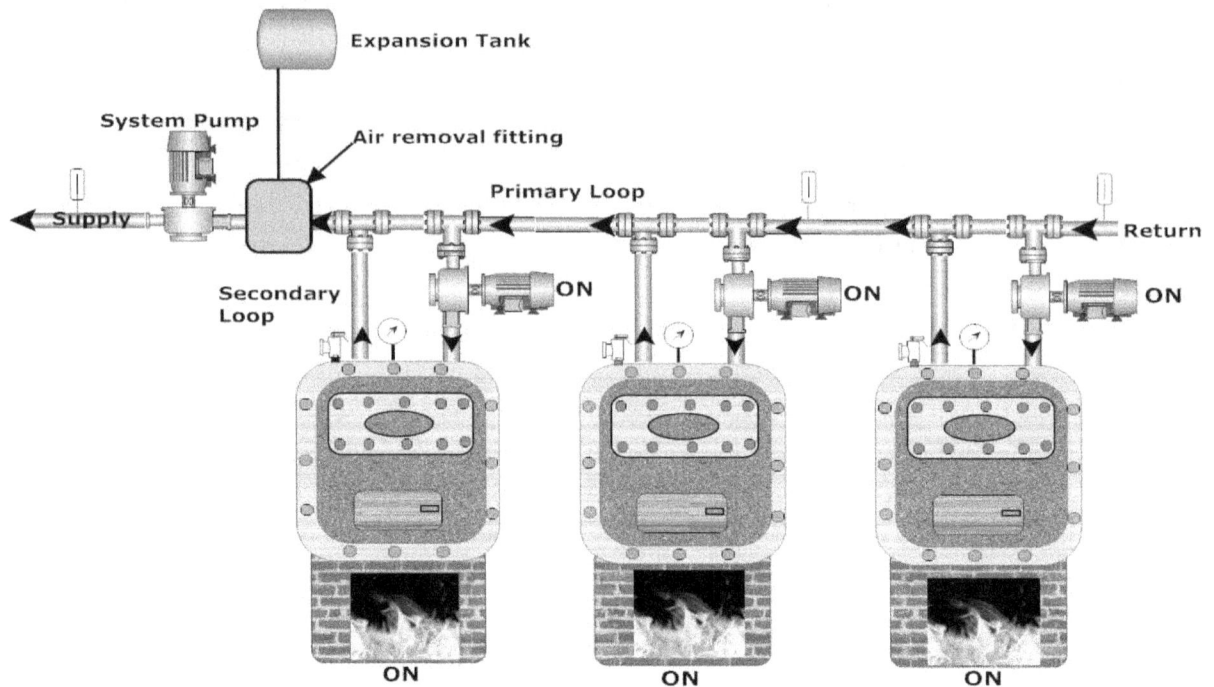

Primary Secondary A Primary Secondary system is a system of hydronic piping loops. The primary loop is the heating loop that delivers heat to the building. It distributes the hot water to the building components. The secondary loop is the piping that pulls water from the primary loop and goes through the boiler and returns back to the loop. The secondary loop could also be used in the system to feed different components but for the purpose of this book, we will refer to the secondary loops as the ones that feed the boilers.

This piping arrangement allows you to meet the requirements of the International Energy Conservation code because it has the ability to isolate the idle boiler when it is not firing. Since each boiler has its own dedicated pump, each boiler is assured that it has the proper flow.

If using primary secondary piping, the takeoffs to the boiler should be within 12" of each other. If they are further apart, you could have ghost flow throw the idle boilers, increasing your operating costs and boiler off time losses.

Manifold Primary Secondary To tweak some more savings from a primary secondary system, my friend Matt Little, Boiler Guru, uses a Manifold Primary Secondary piping. On traditional Primary Secondary systems, the boilers are piped in parallel and also series with one another. For example, if you have three boilers in a row and the upstream boiler starts, it will pull water from the main, heat it and return the warmer water to the main. See drawing on next page. The next downstream boiler will see warmer water than the first boiler. It will be slightly less efficient. This goes with the third boiler as well. If you pipe it with a manifold, each boiler will have access to the colder water, increasing system efficiency.

Standard Primary Secondary Piping

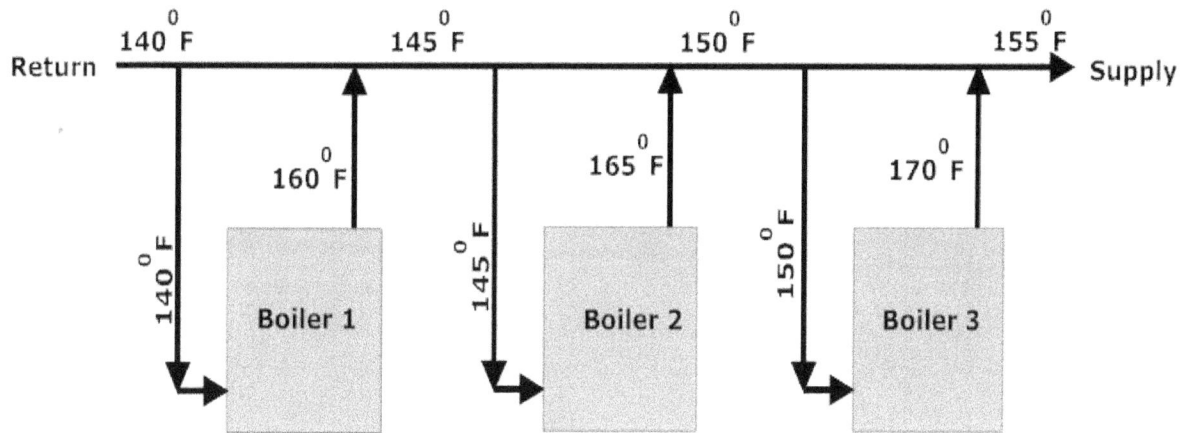

Return 140°F   145°F   150°F   155°F Supply

160°F   165°F   170°F

140°F   145°F   150°F

Boiler 1   Boiler 2   Boiler 3

Manifold Primary Secondary Piping

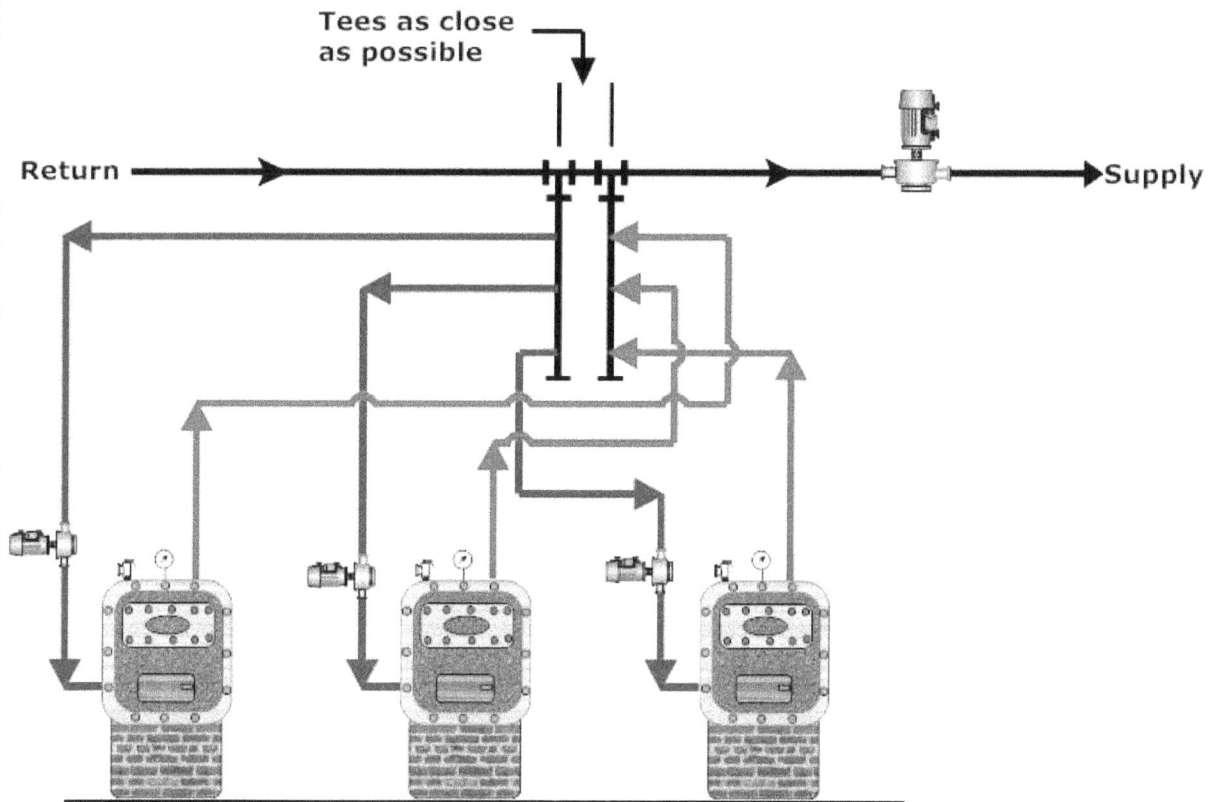

Tees as close as possible

Return   Supply

39

Some thoughts on Primary Secondary Systems There are basically two types of Primary Secondary piping systems, Parallel and Series. Series piping is the most common. There are a couple disadvantages to the series method. As mentioned above, the system efficiency is slightly lower due to the increased water temperature at each downstream boiler. The other disadvantage is that each zone could have a different temperature supply water because it is downstream of the return of the previous zone. See the drawing below that the supply water temperature for Zone 3 is 18 degrees F lower than the supply water temperature to zone 1. To allow access to all the zones, you could pipe it like the manifold primary secondary above or use the parallel method below. The parallel piping arrangement allows the same water temperature to all three zones. It uses crossover bridges to accomplish that. Proper flow connections are crucial in a primary secondary piping system. If the outlet of the boiler is piped upstream of or before the inlet, the boiler will short cycle and the building will not have enough heat.

Series Primary Secondary Piping

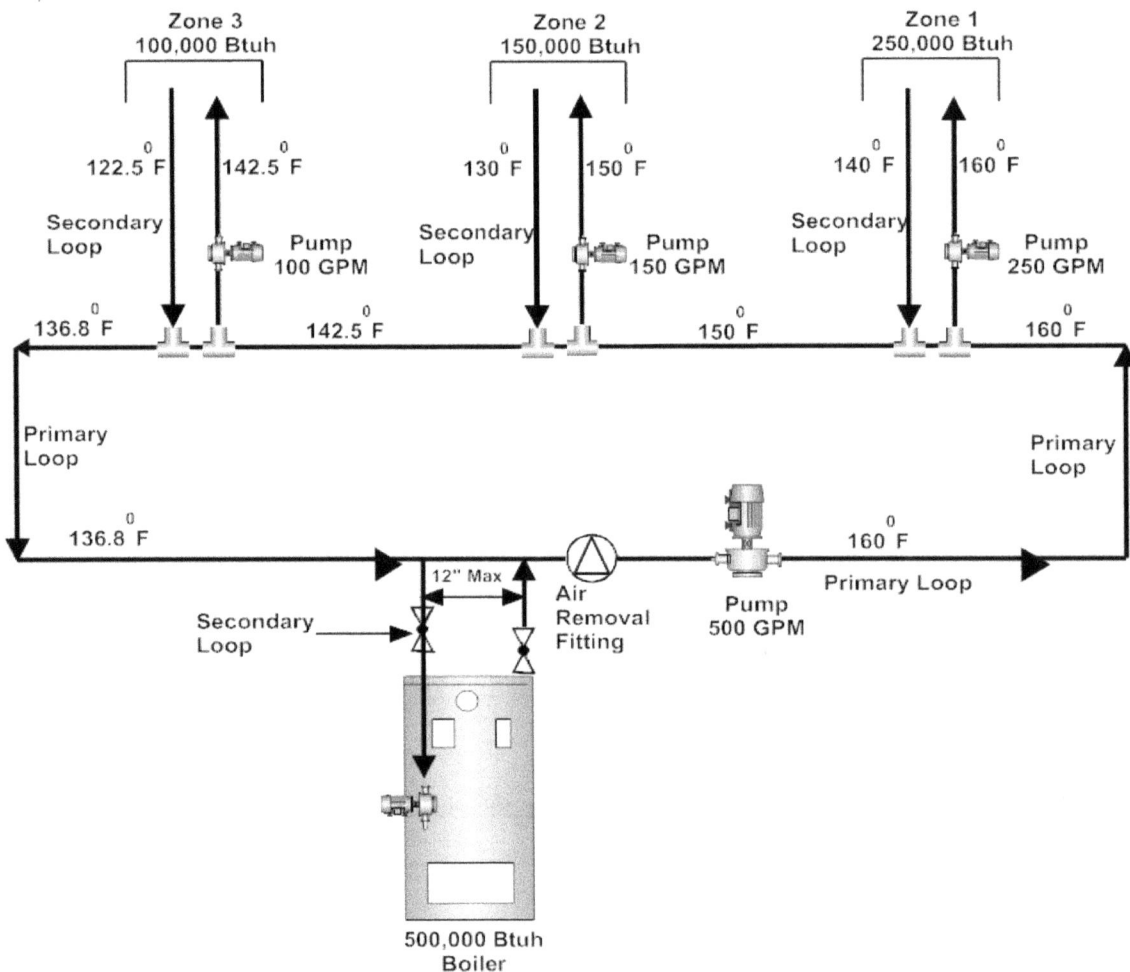

40

Parallel Primary Secondary Piping

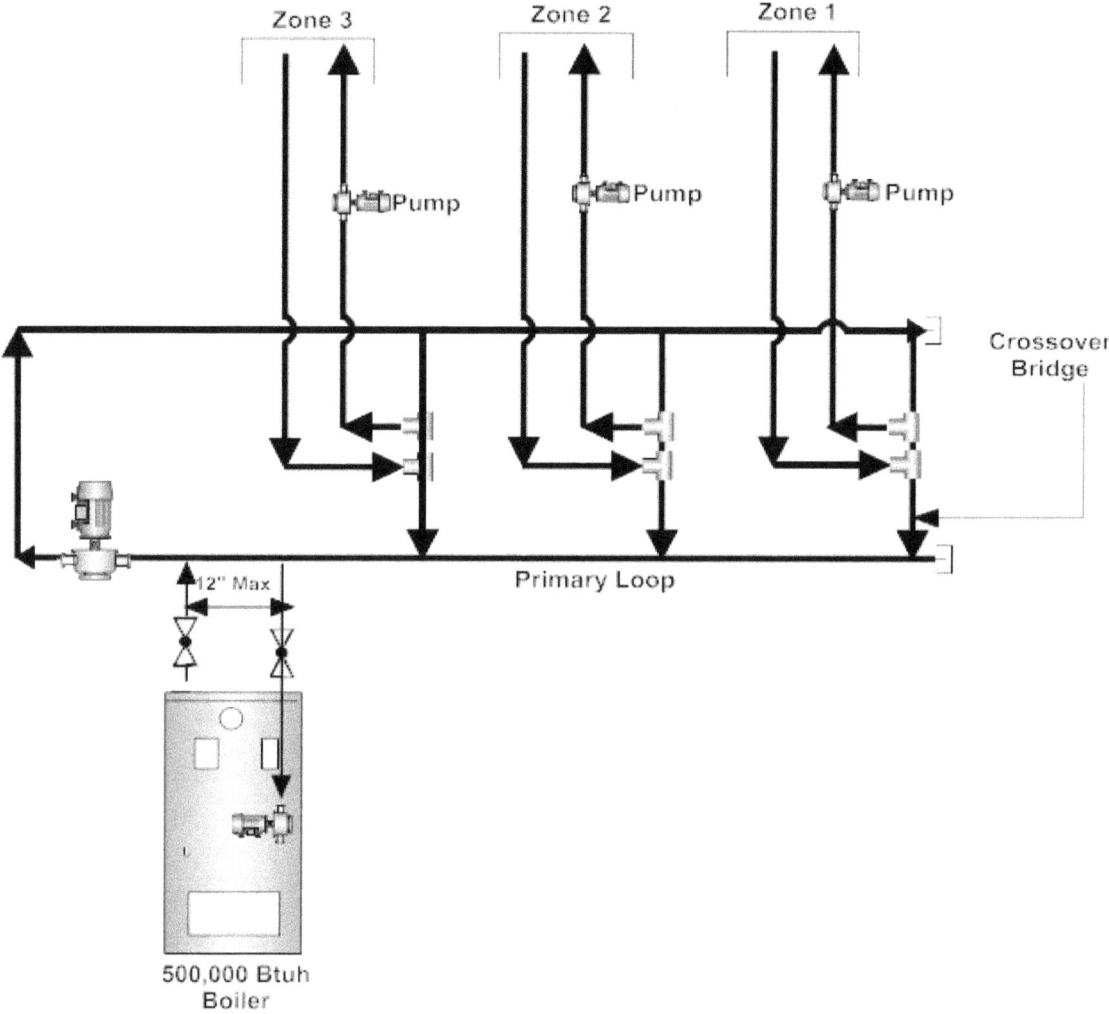

Zone 3   Zone 2   Zone 1

Pump   Pump   Pump

Crossover
Bridge

12" Max   Primary Loop

500,000 Btuh
Boiler

An un-insulated valve will lose
as much heat as a 4 foot
section of un-insulated pipe

Hydraulic Separation A newer method of piping is referred to as Hydraulic Separation or Hydraulic Uncoupling. This is not to be confused with the actress Gwyneth Paltrow describing her divorce as a "Conscious Uncoupling." In this piping method, you do not need a pump on the primary loop so this will save operating and installation costs. The system uses a buffer tank to create the hydraulic separation. The hydraulic separation means that the operation of one pump does not affect the operation of the other system pumps. In a traditional primary secondary system, the hydraulic separation is accomplished by the closely spaced tees that feed the secondary loops. In addition, the buffer tank reduces the short cycling and sooting inherent with low mass boiler systems because it adds volume to the hydronic system. Low mass boilers have a tendency to soot or short cycle under light load conditions. This also reduces the chances of thermal shock to the boiler as the warmer supply water mixes with the cooler return water.

Buffer Tank Courtesy: Boiler Buddy

Pipe Size When visiting the job site, confirm the existing pipe is large enough to handle the required flow of the system. That is why you want to ask about how well the old boiler performed. If the existing system heated the building with pipes that were undersized for the boiler, you may be able to reduce the size of the replacement boiler.

When designing a new system, question everything. I will verify the supply and return pipe size, the pump sizing and GPM, and flue and chimney sizing. It may seem like overkill but it will pay for itself at some point in your career. We visited one project and the existing boiler was in excess of 2,000,000 Btuh. When we checked the existing piping and pump, they were sized for about 900,000 Btuh. The existing boiler did cycle frequently, according to the building contact. If we had sold the client a 2,000,00 Btuh boiler, the pump and piping would have been too small. This could have damaged the new boiler. On another project, we found that the flue was undersized for the heating load required for the building. We had to sell a boiler with direct venting capability through the side wall.

Un-insulated Steam Piping

Pipe insulation According to the 2012 International Energy Conservation Code, all piping serving as part of a heating or cooling system shall be thermally insulated in accordance with Table C403.2.8 The following is list of insulation thicknesses as recommended by the code. When insulating the piping, the fittings should be done as well. Did you know that an un-insulated valve will lose as many Btus as a four foot section of pipe?

Care has to be taken when planning the pipe routing as the insulation of the new pipes has to be factored. For example, if you have two steam pipes within close proximity of each other, you have to consider that the code calls for 3" insulation around each pipe.

If the piping has no insulation, should you include that with the new system? I believe that you should. I think if you educate the owner on the benefits of insulation, they will agree to it.

Minimum Insulation			
Fluid operating temperature			
Pipe Size	105-140	141-200	201-250
Minimum insulation			
<1"	1"	1 1/2"	2 1/2"
1 - 1 1/2"	1"	1 1/2"	2 1/2"
1 1/2 - 4"	1 1/2"	2"	2 1/2"
4 - 8"	1 1/2"	2"	3"
>8"	1 1/2"	2"	3"
Source: 2012 International Energy Conservation Code C403.2.8			

Some Thoughts on Old Gravity Hydronic Systems

Before circulators were invented, some old designers installed a gravity system. This was an ingenious system that relied on the buoyancy of lighter warm water to rise and the cooler denser water would fall. They were installed in buildings up to three stories high. When installing a circulator on these systems, there are a couple items to remember. As a way of balancing the system, an orifice was placed by the original installer in the highest radiator to limit the flow. They are usually found in the union for the shut off valve. They will need to be moved to the first floor.

The second items is that the take offs to the first floor were usually at a 45 degree angle and the take takeoffs to the higher floors were at a 90 degree angle. If the pipe is coming straight of the side of the main header, it was for the top floors.

There is a great book by Dan Holohan that explains these old systems called Classic Hydronics.

Connecting to the Pumps

Multiple pumps were a common way of providing zone control as each zone would have a thermostat that would control the zone pump. There are a couple drawbacks to this system. The first is that you are not pumping away from the compression tank so the inlet of the pump could actually go sub atmospheric or into a vacuum, pulling air into the system through air vents and valve packing. The other drawback is that unless all the pumps are operating, there may be insufficient flow through the boiler which could damage the boiler.

Verify that the circulator is properly sized for both the heat loss of the building and the heating system. Some boilers require a minimum flow. Confirm the flow requirements of the boilers.

When replacing the pumps on the project, be sure to talk with the pump supplier for other items you may require. A common

component we see on the outlet of the pump is a **triple duty valve**. This valve is a combination of check valve, balance valve and shutoff valve.

Triple Duty Valve

45

Another common accessory for a circulator is a **suction diffuser**. It is on the inlet to the pump. It has a strainer to limit the dirt that may flow into the pump. It also has straightening vanes to reduce noise and the stress on the impeller due to uneven flow.

Isolation Valves could cause some serious damage if closed when the pump is running. This is called Dead Heading the pump. If the valves are closed, the temperature and pressure inside the pump housing rises to a dangerous level very quickly and could damage the pump or could even explode. Consider that if a small B& G Series 100 pump with a 1/12 HP motor is running with closed valves, the temperature inside the circulator will rise 50^0F per hour.

Expansion joint Some pumps require an expansion joint to absorb the vibration inherent in a moving pump. Be careful that the pump anchors actually allow the expansion joint to capture the vibration.

Pressure gauges Provisions should be made for pressure gauges on the inlet and discharge of the pumps. This will help to both adjust and trouble shoot the pump.

Pump Mounting Options

According to ASHRAE, a base mounted pump has a life expectancy of 20 years while a pipe mounted pump will last only 10. The following show the different mounting arrangements.

Base Mounted Pump

Pipe Mounted Pump

Sizing a Circulator

There are a couple short cuts to sizing a pump for a boiler. Most boilers are designed for a 20 to 30 degree rise through the boiler. To size a pump for a boiler and maintain a 20 degree rise through the boiler, divide the output of the boiler by 10,000 to get the proper GPM for a 20 degree rise.

If the boiler can handle a 30 degree rise, divide the output of the boiler by 15,000 to get the proper GPM or Gallons per Minute.

Example: To see if the existing 40 GPM pump on a project is large enough for the new boiler, let us look at the equipment. Our new boiler has a rated output of 800,000 with a design temperature rise of 20 degrees F. The existing pump is 40 GPM. If we divide the boiler output by 10,000, we see that the boiler will require an 80 GPM pump. This is double the GPM of the existing pump. Our flow would be half and the temperature rise would be double, possibly ruining the new boiler. In this case, we would have to replace the pump. If our new boiler can handle a 30 degree rise, we could divide it by 15,000 and find that the boiler will require a 53 GPM pump. The existing pump is still too small for this boiler.

If you would like to see how I arrived at the 10,000 or 15,000 number, the following is the formula:

$$\text{GPM} = \frac{\text{Rated output of boiler}}{8.33 * 60 * \triangle\,°F}$$

or

$$\text{GPM} = \frac{\text{Rated output of boiler}}{500 * \triangle\,°F}$$

*500 = 8.33 * 60*

GPM = Gallons per minute flow rate
8.33 = Weight of a gallon of water
60 = Converts the formula from hours to minutes aka Gallons per Minute GPM
$\triangle$ °F Temperature rise through boiler is usually about 20-30 degrees F.

The following is the actual formula for a 20 degree rise for the 800,000 Btuh boiler:

$$\text{GPM} = \frac{800,000\ Btuh\ (Output\ of\ boiler)}{8.33 * 60 * Temperature\ rise\ through\ boiler}$$

$$\text{GPM} = \frac{800,000\ Btuh\ (Output\ of\ boiler)}{500 * 20\ Degree\ rise}$$

*500 = 8.33 * 60*

48

$$80 \text{ GPM} = \frac{800,000 \; Btuh \; (Output \; of \; boiler)}{10,000}$$

The following is the actual formula for a 30 degree rise for the 800,000 Btuh boiler:

$$GPM = \frac{800,000 \; Btuh \; (Output \; of \; boiler)}{8.33*60*Temperature \; rise \; through \; boiler}$$

$$GPM = \frac{800,000 \; Btuh \; (Output \; of \; boiler)}{500*30 \; Degree \; rise}$$

$$53 \text{ GPM} = \frac{800,000 \; Btuh \; (Output \; of \; boiler)}{15,000}$$

The following is a chart to help you verify the pump size based upon common delta T

Boiler Output	Temp rise through boiler		Boiler Output	Temp rise through boiler	
	20⁰F	*30⁰F*		*20⁰F*	*30⁰F*
	GPM	*GPM*		*GPM*	*GPM*
500,0000	50	33	1,500,000	150	100
600,000	60	40	1,750,000	175	117
700,000	70	47	2,000,000	200	133
800,000	80	53	2,500,000	250	167
900,000	90	60	3,000,000	300	200
1,000,000	100	67	4,000,000	400	267
1,250,000	125	83			

49

Pumping away

The location of the circulator to the system is crucial to the proper operation of the system. Many of the older boilers used to have the pumps on the return piping. It was rumored that this was because the seals inside the pumps could not handle the higher temperatures on the supply piping. I believe that it made it easier to package the boiler as the pump would not be above the crate making itself a target for a fork lift driver. The drawback to having the pump on the return is that it could bring air into the system. The pump creates flow by creating a pressure differential in the piping. The discharge of the pump will be at a higher pressure than the inlet. If both pressures are the same, the system will have no flow. So, if we have the pump on the return, the discharge of the pump will be facing the boiler and the compression tank. When the pump turns on, the pump discharge pressure is negated by the compression tank. The extra pressure is absorbed by the compression tank. The pump does not realize that so it will try to create a pressure differential in any way that it can. If it cannot increase the discharge pressure, it will decrease the suction. In some applications, this can actually pull the inlet into sub atmospheric condition or a vacuum. If this happens, air will enter the piping. This could cause corrosion inside the boilers and pipes. It will also create comfort complaints as the air will find its way to the top floor and you will have air bound radiators that have to be bled. In severe applications, you may have to bleed the radiators several times a year.

The preferred location for the pump is on the discharge of the compression tank. This is referred to as "Pumping Away." When the pump starts it will increase the discharge pressure on the outlet and the inlet stays at the pressure setting of the water feeder. In this way, the system will be less susceptible to internal corrosion and air bound heat emitters.

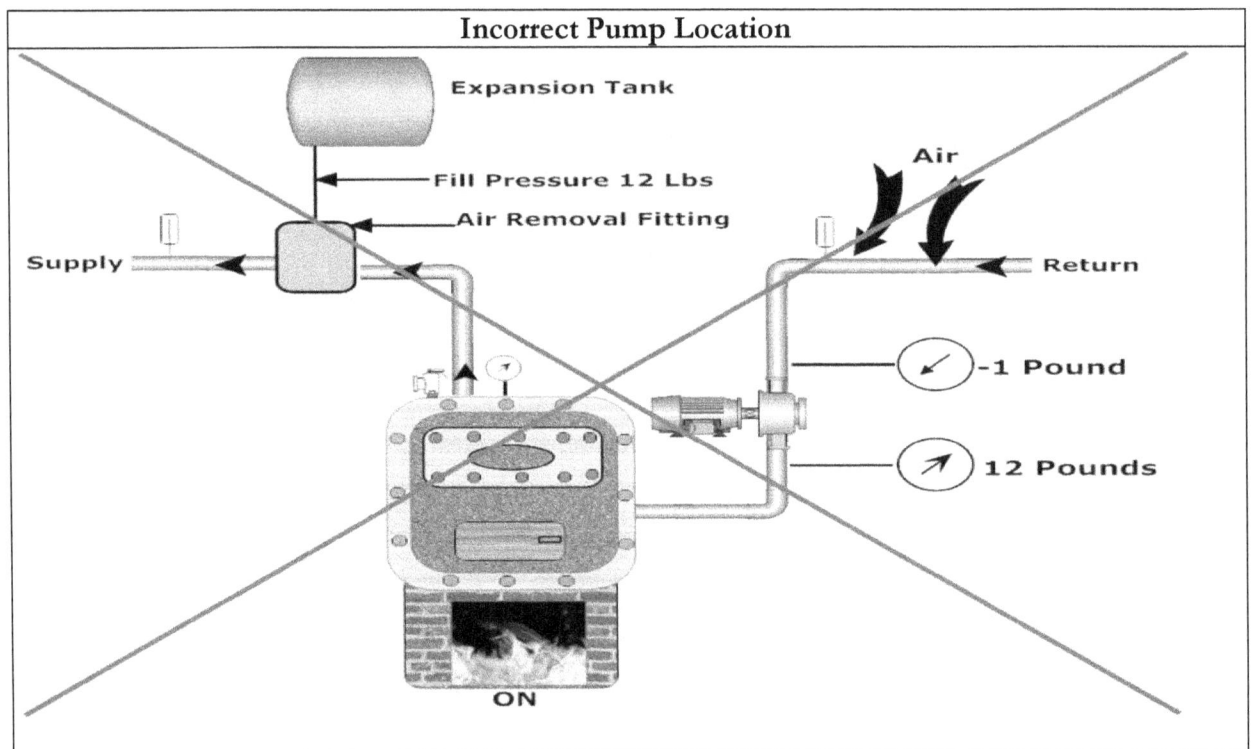

Incorrect Pump Location

Expansion Tank

Fill Pressure 12 Lbs

Air Removal Fitting

Air

Supply

Return

-1 Pound

12 Pounds

ON

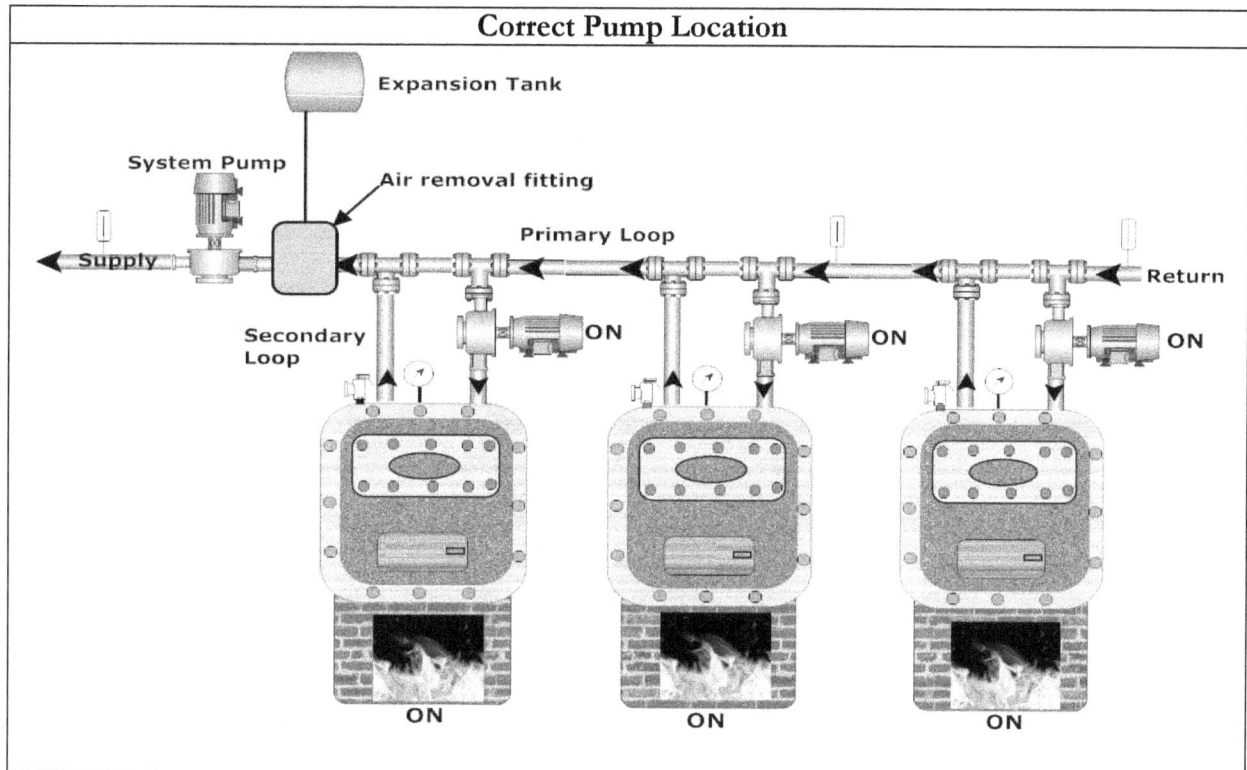

Correct Pump Location

Calculate pump head

1 Measure longest run in feet. Include both supply and return.

2 Multiply by 1.5 to calculate fittings and valves

3 Multiply by 0.04 (4' head for each 100' of pipe ensures quiet operation)

For example, 1,000 feet is longest run

1,000 x 1.5 x .04 = 60 feet of head

or

Measure longest run, divide by 100 feet, multiply by 6 feet to get pump head. 1000 ÷ 100 = 10

10 x 6 = 60 feet

Pressures gauges are a good idea for both the supply and return of a pump. It will let you know the pressure available for the suction of the pump as well as the pump discharge pressure. The difference between the two pressures is the ability of the pump to overcome the resistance of the piping. For example, our system above requires 60 feet of head. To see what pressure we need, divide the feet required by 2.31 and we would need a pressure differential of 25.9 psi. You would have to look at the pump curve to see if the pump can furnish the required GPM at the pressure of 25.9.

Who are you gonna call? When you have two or more loops each with its own circulator, you could experience ghost flow through the idle zones. For example, if loop B is running, you could have ghost flow backwards through Loop A. This could lead to overheating the idle zone that Loop A handles. To eliminate the chance of that happening, flow control valves are often used. A flow control valve is like a weighted check valve. It is only opened when the zone pump for that zone is operating.

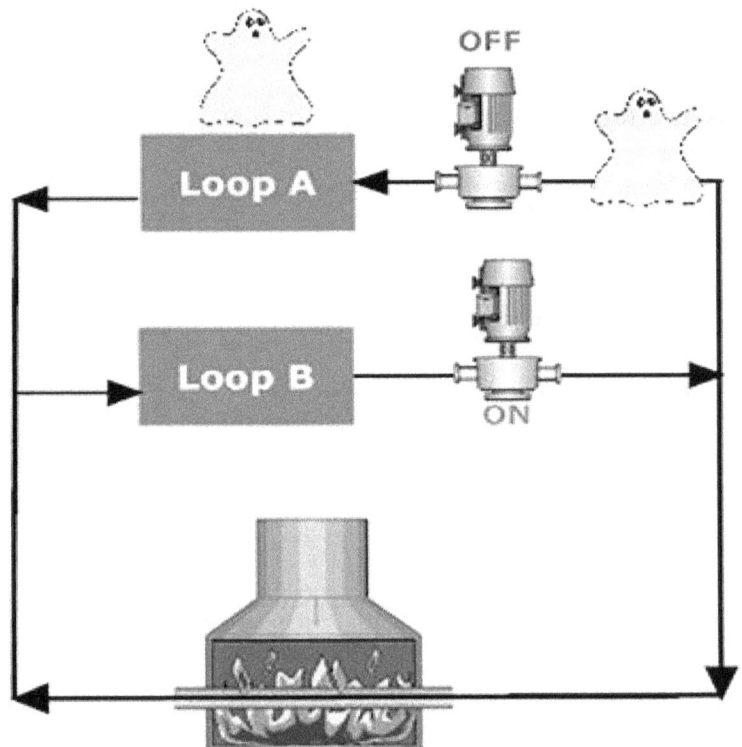

Connecting to the Hydronic Accessories

Relief Valve The relieving capacity of the relief valve should be rated to be able to fully relieve the pressure in the boiler. The relief valve discharge pipe has to be the same diameter and cannot be reduced. If the outlet pipe is reduced, the relieving capacity of the relief valve is lower. Be careful about piping the discharge of the relief valve hard to the ground. On a local project, the discharge piping was piped solid to the ground using copper tubing. When the relief valve opened, the temperature of the water caused the tubing to expand. This expansion stressed the brass relief valve and cracked it, flooding the boiler room. Many relief valve manufacturers recommend using schedule 40 pipe or not allowing the discharge piping to be piped solid to the ground.

The relief valve on steam systems greater than 500,000 Btuh should be vented outside.

If you want to use copper, I would suggest using an air space similar to the picture below. If the relief valve opens, the pipe expansion would not stress the relief valve. The discharge piping has to be supported so the weight of the piping is not all on the relief valve.

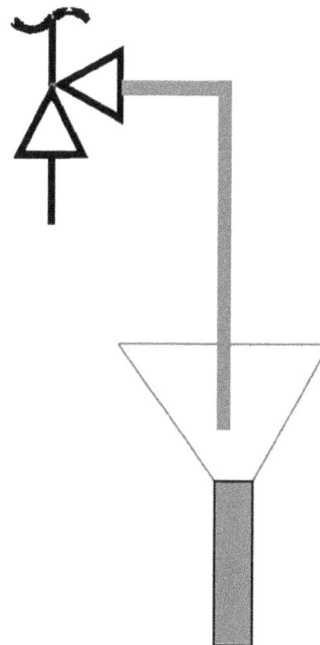

Relief Valve piped solid to ground	Relief Valve piped solid to ground

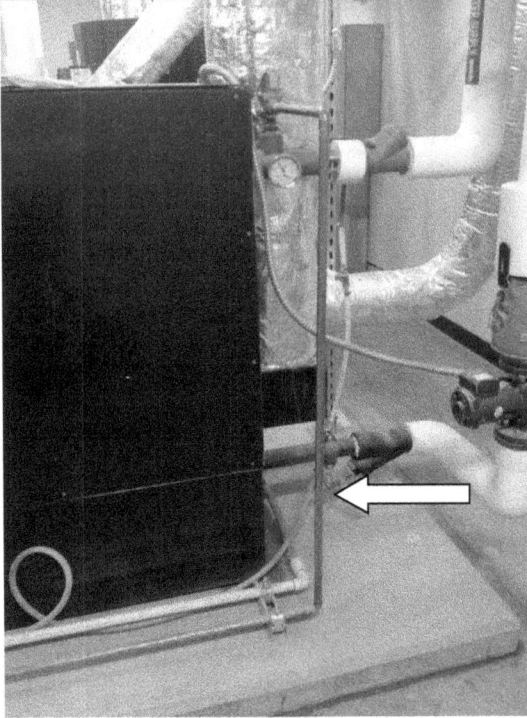

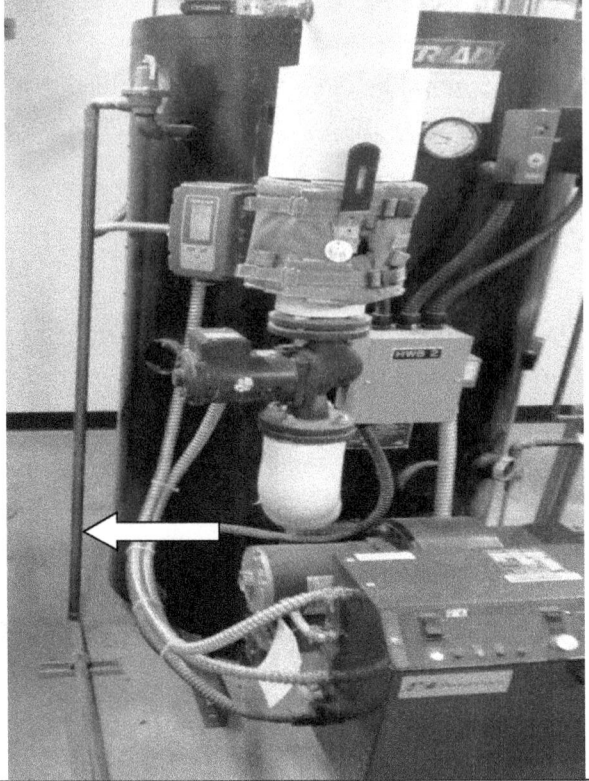

Ray's Rule #5 Some Clients Lie.

Compression or Expansion tanks

When water is heated from 65^0F to 180^0F, it will expand at 3% of its volume. This expansion of the water has to be controlled or managed. This is typically done with a compression tank. Although they are frequently referred to as either a compression or expansion tank, there is a difference. The expansion tank was used on old gravity hydronic systems. It was located in the highest part of the building and was open to atmosphere. When water was heated, it expanded and the level rose in the open tank. The compression tank is located in the boiler room and is a cylinder which has an air cushion on the top of the tank. When water is heated, the expansion compresses the air inside the tank. When the water cools and the system pressure reduces, the air cushion is greater. When looking at the gauge glass on the compression tank, it should be about one quarter to halfway up when the water is cool and about 3/4 of the way up the gauge glass when the water is warm. Without this tank, the expansion could cause the water to leak through the valve packing or lift the boiler relief valve. You may see the relief valve discharge piping weeping when the tank is flooded.

If the compression tank has an internal diaphragm, verify its integrity and the internal pressure is correct. The air pressure is checked with a fitting similar to one on your vehicle tire. As a matter of fact, you use a tire pressure gauge to check the pressure. It is like an inner tube in a tire. The air pressure should be checked while it is empty and not connected to the piping.

If you are re-using the existing compression tanks, the sizing should be verified. *There are formulas in the rear of this book that allow you to verify the size of the tank.* The tanks should be inspected for leaks. Look at the top of the tank for pin holes as this will allow the tank to flood. A way to test if the compression tank is flooded or the piping is plugged to the tank is to watch the boiler PTA or Pressure Temperature Altitude gauge as the boiler fires. If the pressure climbs when the water heats, this usually means that the tank is flooded or the piping to the tank is plugged.

Airtrol fitting

Check the Airtrol fitting for the tank as well. The Airtrol fitting has a long tube that will go about 2/3 of the way up the tank. This allows the air to be added to the top of the tank. Perhaps you remember the game you played as a youth with a straw, holding your thumb over the top and pulling it from the drink. The liquid stayed in the straw until you removed your thumb. That is what happens with the compression tank when the Airtrol fitting is not working.

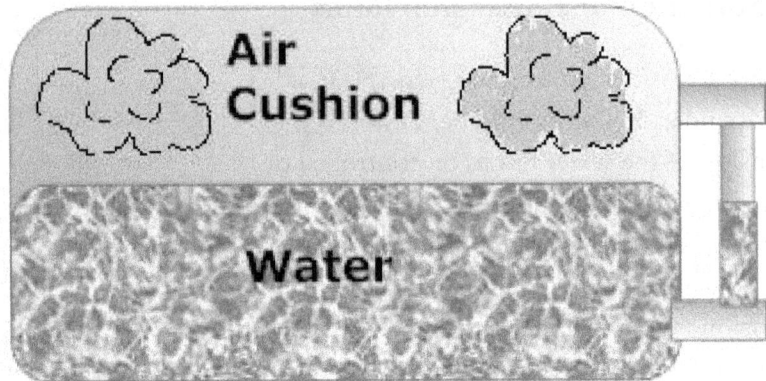

If you have verified that the tank is large enough and plan to re-use the tank, I have several suggestions. The first is to replace the gauge glass fittings on the tank. The packing on the valve dries out and this will allow the air to leak out, resulting in a flooded tank. The Airtrol fitting should also be replaced as the vent tube sometimes breaks. I would also suggest checking the piping from the hydronic loop to the tanks. It has a tendency to plug and cause the relief valve to open or weep.

The following is the expansion in inches of different piping:

Thermal Expansion of Piping Material in inches per 100 feet from 32 Deg F

Temperature Deg F	Carbon & Carbon Moly Steel	Cast Iron	Copper
32	0	0	0
100	0.5	0.5	0.8
150	0.8	0.8	1.4
200	1.2	1.2	2.0
250	1.7	1.5	2.7
300	2.0	1.9	3.3
350	2.5	2.3	4.0
400	2.9	2.7	4.7
500	3.8	3.5	6.0
600	4.8	4.4	7.4
700	5.9	5.3	9.0

Horizontal Compression Tank	Vertical Compression Tank

Ray's Rule #2 A piece of equipment will fail 5 minutes after the supply house closes.

Rule of Thumb for Compression Tank Sizing Steel Piping

Based upon the following:
Entering Pressure 10 pounds, Maximum Pressure 25 pounds
Entering Temperature 40°F, Maximum Temperature 220°F

Steel Piping

System Capacity in Gallons	Closed Compression Tank	Diaphragm Tank	System Capacity in Gallons	Closed Compression Tank	Diaphragm Tank
200	39	23	1,700	327	195
300	58	34	1,800	347	206
400	77	46	1,900	366	218
500	96	57	2,000	385	229
600	116	69	2,500	481	287
700	135	80	3,000	578	344
800	154	92	3,500	674	401
900	173	103	4,000	770	458
1,000	193	115	4,500	867	516
1,100	212	126	5,000	963	573
1,200	231	138	6,000	1,156	688
1,300	250	149	7,000	1,348	802
1,400	270	160	8,000	1,541	917
1,500	289	172	9,000	1,733	1,032
1,600	308	183	10,000	1,926	1,146

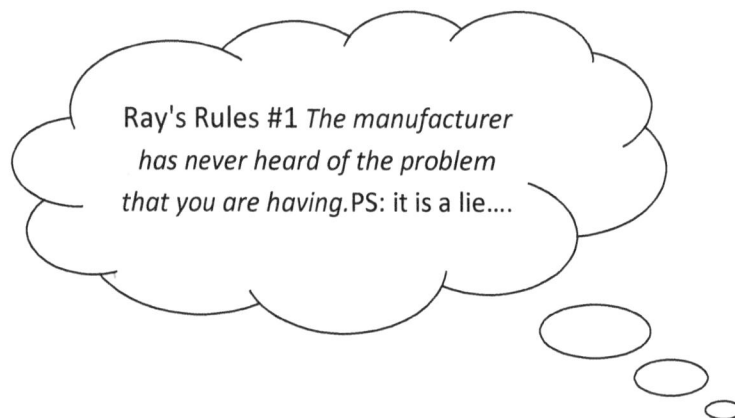

Ray's Rules #1 *The manufacturer has never heard of the problem that you are having.*PS: it is a lie....

Rule of Thumb for Compression Tank Sizing Copper Piping

Based upon the following:
Entering Pressure 10 pounds, Maximum Pressure 25 pounds
Entering Temperature 40°F, Maximum Temperature 220°F

Copper Piping

System Capacity in Gallons	Closed Compression Tank	Diaphragm Tank	System Capacity in Gallons	Closed Compression Tank	Diaphragm Tank
200	37	22	1,700	315	188
300	56	33	1,800	334	199
400	74	44	1,900	352	210
500	93	55	2,000	371	221
600	111	66	2,500	463	276
700	130	77	3,000	556	331
800	148	88	3,500	649	386
900	167	99	4,000	742	441
1,000	185	110	4,500	834	496
1,100	204	121	5,000	927	552
1,200	222	132	6,000	1,112	662
1,300	241	143	7,000	1,298	772
1,400	260	154	8,000	1,483	883
1,500	278	165	9,000	1,668	993
1,600	297	177	10,000	1,854	1,103

Multiple Compression Tanks Rules of thumb

- The vertical pipe between air separator and tanks should be 3/4" or larger.
- The horizontal pipe to the manifold tanks should be as follows:
- 3/4" if pipe run is less than 7 feet long.
- 1" if pipe run is 7 feet to 20 feet long.
- 1 1/4" if pipe run is 21-40 feet long.
- 1 1/2" if pipe run is between 40-100 feet long.

Tank Manifold Pipe Sizing

- 1" for 2 tanks.
- 1 1/4" for 3-4 tanks.
- 1 1/2" for 5 or more tanks.

Backflow Preventer

Even though the system may have a backflow preventer on the water supply to the building, you will need a backflow preventer on the water feed from the building to the boiler. A backflow preventer is a sophisticated check valve that assures the water on the outlet will not mix with the water on the inlet of the device. A backflow preventer should be installed on the feed water pipe to the boiler. You do not want that nasty black water, typically found inside the boiler, to be mixed into the building's drinking or potable water supply. A health care facility in this area had the boiler water connected to the potable water on a new addition. It was never detected until the facilities manager changed water treatment companies and the new water treatment specialist recommended a dye in their chemicals to show if it was leaking. Shortly afterward, a nurse filled a white Styrofoam cup with water and noticed a pink hue to the water. She called the maintenance department and they found the cross connection. It was an ugly scene and lawyers were everywhere.

When installing the backflow preventer, it should be installed upstream or before the bypass around the feeder. The picture shows a backflow preventer, water feeder and bypass. Please pardon the crooked piping. No, I did not install it.

Water transfers heat 3,500 times more efficiently than air.

Water Feeder

How do you get the water to the highest radiator in a building? Pump? Reverse Gravity Device? or perhaps Yoda from Star Wars? Many people believe that it is the pump or circulator that pushes the water to the top of the building. Instead, it is actually system pressure that raises the water to the highest point. That job falls to the water feeder or PRV, pressure reducing valve. How much pressure do you need? Well, one pound of pressure will raise water 2.3 feet. A good rule of thumb for tall buildings is that the boiler pressure should be half the height of the highest radiator. If the highest radiator is 50 feet high, the system fill pressure should be 25 pounds.

Why are some boilers in the penthouse? As I said earlier, one pound of pressure will raise water 2.3 feet. Conversely, a column of water has a weight of that same amount. If you have a tall building, the weight of the water at the bottom could be substantial. If we have a twenty story high rise building and the highest radiator is at the top floor 300 feet above the boiler, the weight of water would be around 130 pounds at the bottom of the building. Some very tall buildings have mechanical rooms at different floors because of the pressure exerted from the column of water.

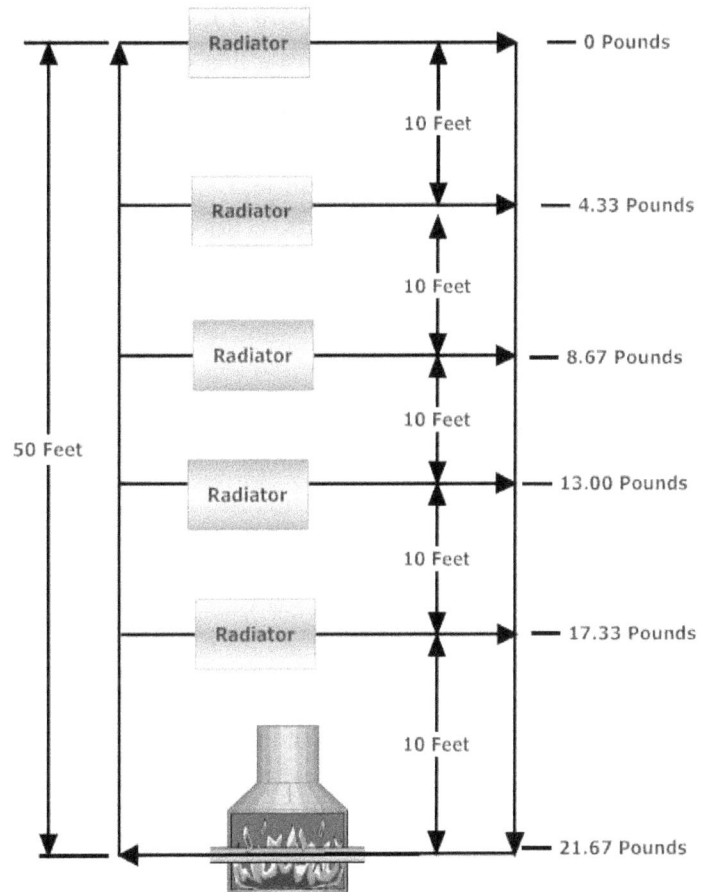

Expansion Tank

Bypass

Backflow Preventer

City Water

PRV

Primary Loop

Return

To Boiler

Air Removal Fitting

Supply

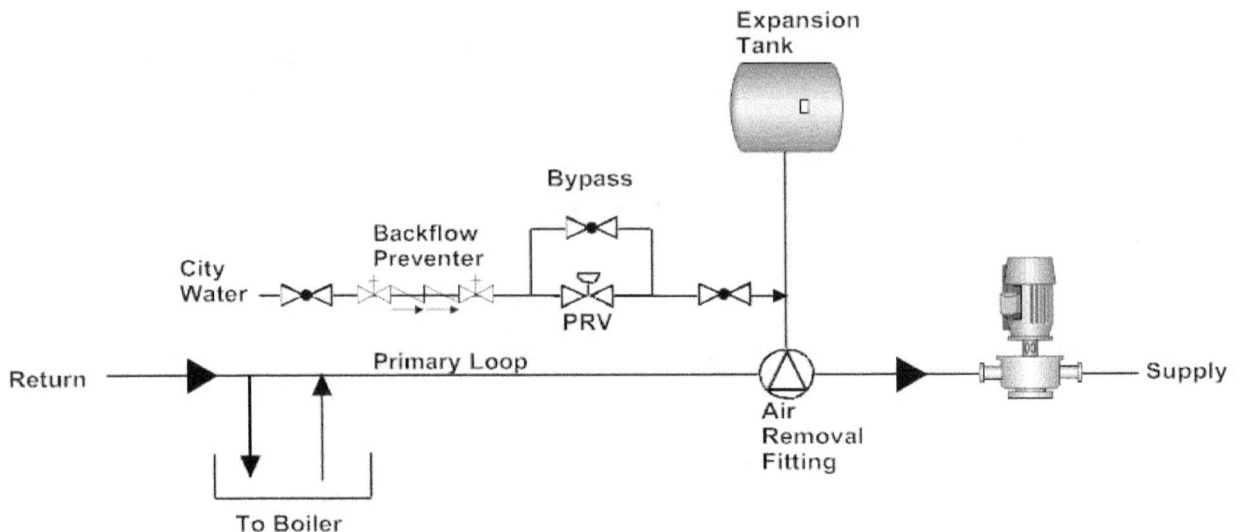

The residential feed valves are usually adjustable from 10 - 25 psi. The pressure reducing valve is usually set at 12 pounds of pressure at the factory. This pressure was used to assure that water would reach the top radiator of a traditional two story home. On commercial systems, it may require more pressure to get it to the highest radiator. One pound of pressure will raise water 2.3 feet. If the highest radiator is on the top floor of a five story building, it may be 50 feet high. To see how much pressure we need for that elevation, divide the height of the highest radiator by 2.3 feet. In this example, it would be 50 feet divided by 2.3 psi = 21.7 psi or round it up to 22 psi. That just assures that we have the water to that elevation, Bell and Gossett suggests that you add an additional 3-4 psi to allow the air to vent and the water to not flash to steam inside the pipe. That will take us to between 25 - 26 psi. That may be at the far end of our water feeder rating and will need a larger one. I suggest replacing the feeder on replacement boiler projects. They could come back to bite you later.

Feed valve piping is important in a hydronic system. A bypass loop around the feeder should be installed to allow quick fill of the system. You will thank me when filling the commercial system by using only the fill valve. Make sure that the bypass piping does not also bypass the backflow preventer. A boiler inspector made us re-pipe our bypass loop as it bypassed the backflow and the feed valve. He felt that the bypass could be opened and the dirty boiler water could be fed back to the building's potable water supply.

I would also suggest using new valves and unions to allow repair or replacement of the feed valve and backflow preventer without draining the system.

Hot Water System Makeup: Minimum connection size shall be 10% of largest system pipe or 1", whichever is greater 20" pipe should equal a 2" connection.

The following chart shows how much pressure correlates with height in feet of water.

Water Pressure to Feet Head			
Pounds Per Sq Inch	Feet Head	**Pounds Per Sq Inch**	Feet Head
1	2.31	100	230.90
2	4.62	110	253.98
3	6.93	120	277.07
4	9.24	130	300.16
5	11.54	140	323.25
6	13.85	150	346.34
7	16.16	160	369.43
8	18.47	170	392.52
9	20.78	180	415.61
10	23.09	200	461.78
15	34.63	250	577.24
20	46.18	300	692.69
25	57.72	350	808.13
30	69.27	400	922.58
40	92.36	500	1,154.48
50	115.45	600	1,385.39
60	138.54	700	1,616.3
70	161.63	800	1,847.2
80	184.72	900	2,078.1
90	207.81	1,000	2,309.00

Rules of Thumb to Estimate System Volume

In some instances, you may need to estimate the water volume in the system. This is used for water treatment and adding glycol. The following are some rules of thumb to estimate the system volume.

- Multiply steel compression tank volume by 5.
- 35 – 50 gallons per Boiler HP
- Pump GPM x 4
- Compression Tank volume is 20% of system volume
- Rated tonnage of system x 10 gallons

Air Removal

With any hydronic system, air must be removed before hot water could enter the terminal equipment or piping. Check the existing piping or terminal equipment to see how the air could be removed. See whether there are vents in the piping or equipment. Some have automatic vents while most have manual vents. Many of the manual vents use a special key called the radiator vent key that is available in most heating supply houses and big box stores. I like to give the client a few extra keys so they could remove the air from the system. If the system has automatic vents, I would urge you to replace those as they sometimes plug and become inoperable. It is much easier to replace the vent when the system is drained than when heat is required for the building.

Air is the enemy of any hydronic system. It causes internal corrosion of the pipes and boilers. It could also impede the flow of water to areas of the building resulting in comfort complaints. To eliminate air from the system, an air removal fitting is usually used. This device will capture the air and either send it to the compression tank or vent it to the boiler room. The air removal fitting should be installed after boilers and before the system pump. The pipe to the compression tank is also where the makeup water connection for the system is. Air removal fittings work best when a straight piece of pipe 18" or longer precedes the fitting.

One of the most overlooked parts of a replacement project is the time it takes for the air removal. On a large commercial building, it could take several frustrating days to completely remove the air from the system once it has been drained. This has to be included in the price.

A simple way to vent air from the system is to install a ball valve at the top of each riser. You open the valve when you want to vent the air. Install a pipe plug in the outlet of the valve in case the valve leaks when done venting.

Purge fitting On smaller systems, it is a good idea to install a purge fitting in the boiler room on the return piping. It is simply a hose bib that allows the air to go through the system and be vented in the boiler room. Although this does not eliminate all the air, it does rid the system of a great majority. Try to imagine how long it would take to vent a large building of the air using an 1/8" air vent in the top. For this to work, you close valves B and C. Connect hose to purge fitting. Open valve A and allow water to run through the hose until all the air bubbles are gone. You could then open valves B and C. To get rid of air inside the boiler, you may have to open the relief valve.

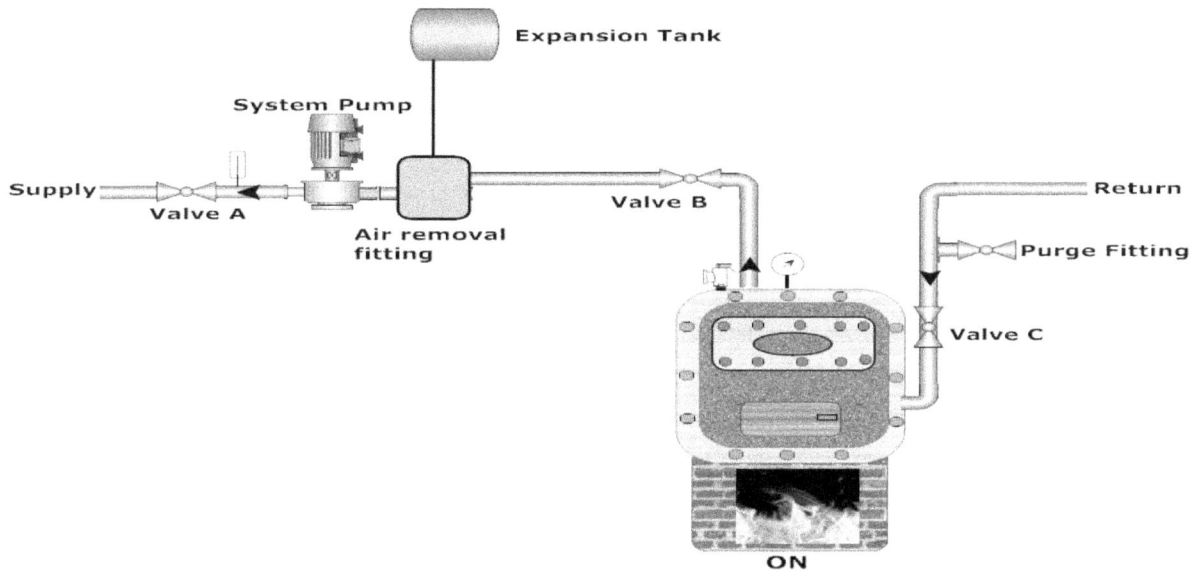

Getting rid of the stubborn air pocket. A certain wing in a building was having a difficult time heating. When we opened the vents, the system spilled water from them. The maintenance department had repaired a leaking coil and could not get heat to another zone after the repair. We correctly diagnosed that the system had an air blockage to the zone somewhere in the system. The challenge was to get the air blockage moved and removed. To do so, we had to perform a series of tasks that finally worked. We would cycle the pumps on and off. This allowed the natural buoyancy of the air to raise to the top of the system. We also raised the system pressure to try moving this large air bubble inside the piping to an area where it could be removed. We would close different zones to force more flow through the blocked ones. It finally moved the air pocket to a place where we could remove it. Another way to rid the system of air is to install about an ounce of Dawn dish detergent. This breaks the air bubbles into smaller units and they can be removed easier.

Air Purge This is another variation on the purge fitting. It will allow you to vent most of the air from your system without having air vents installed in the piping. Here is how it works: Close Valve B. Open Valve C and Valve A. Connect hose to boiler drain 1 on the run of the tee. Open boiler drain 1. Fill your system through the fill bypass. Put the hose into a bucket. Run water until the bubbles stop coming through the water in the bucket.

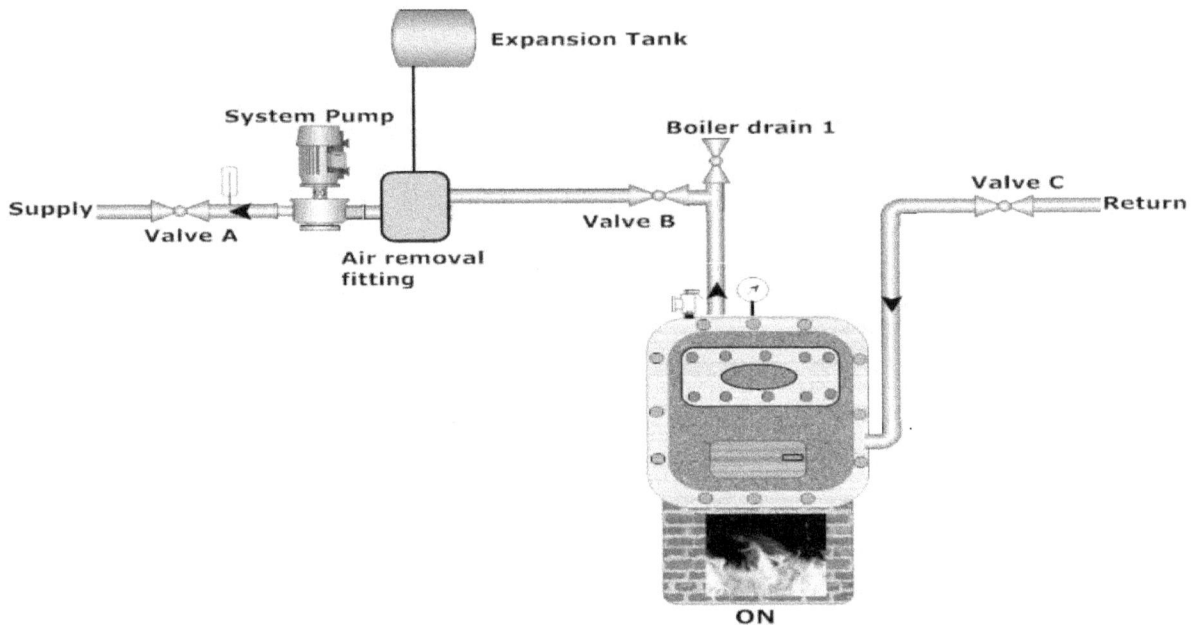

A leaking 1/4" fitting from a gas train can fill a 10 cubic foot room with combustible gas within an hour.

Connecting to the Terminal Equipment

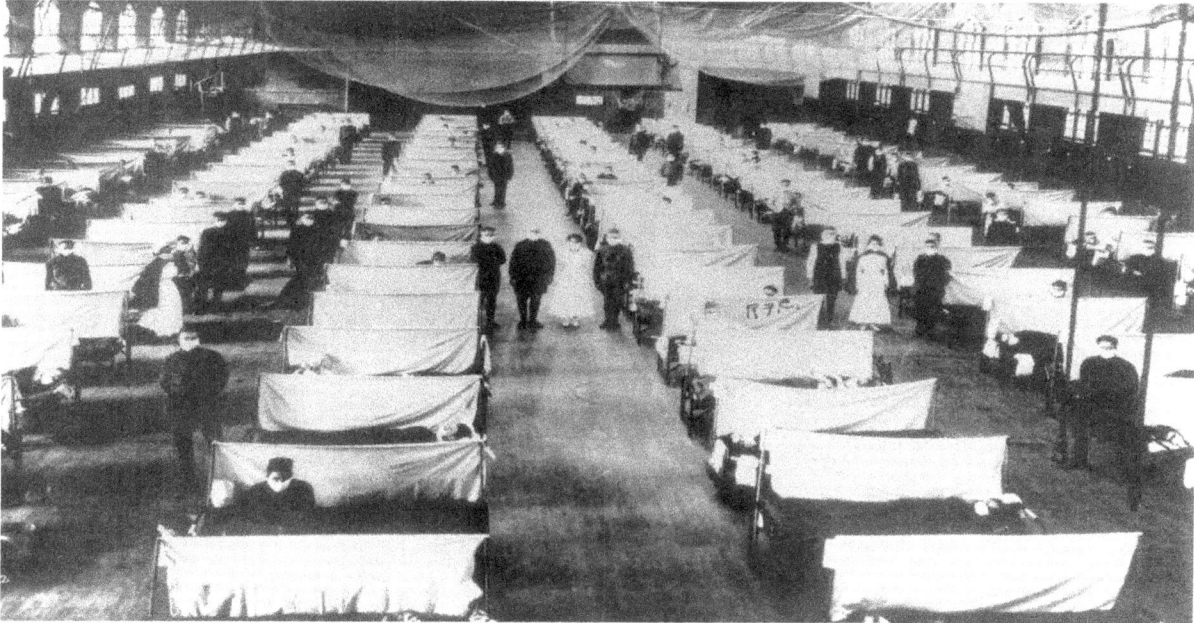

Credit: Office of the Public Health Service Historian

How does a flu pandemic affect the design of heating systems? In 1889, it took just four months for the Russian flu to circle the globe, killing nearly one million people worldwide. In 1918, the Spanish flu decimated the world by infecting between 20-40% of the world's population. It was estimated that between 20-50 million people died from the flu worldwide, surpassing the 16 million that died in World War I. This pandemic caused 675,000 deaths in the U.S. alone. The flu was so deadly that people that were infected in the morning died by nightfall with 20-50 year olds as the most common victims. Since most victims died during the night, it was thought that sleeping with the windows open would avoid the disease. According to a pamphlet entitled, "Spanish Influenza Three-Day Fever The Flu" by the United State Public Health

SPANISH INFLUENZA
PRECAUTIONS

1. Keep in mind that like most contagious diseases influenza is spread by contact, that is, by the transfer of the poison from one person to another. It is spread by sneezing, coughing and spitting at which times the discharges from the nose and throat are scattered in the air.

2. Avoid crowds as much as possible, including moving picture places, theaters and other assembly halls. When feasible avoid crowded street cars.

3. When sneezing or coughing, place your handkerchief before your nose and mouth.

4. Make sure that you are properly clothed, in accordance with the varying changes in temperature, prevalent at this time of the year.

5. Fresh air is always good. Keep your bed room windows wide open, and secure as much sleep as possible.

6. Keep the digestive organs in good condition.

7. Drink water freely.

8. Avoid common drinking cups, common towels and similar utensils.

9. Wash your hands frequently.

10. Use a mild antiseptic as a nose spray or as a mouth gargle, especially if your throat is sore or there is tendency to sneezing.

11. If you have a "cold" use utensils for your personal use exclusively, or if you are in contact with one so affected be careful not to handle utensils used by them.

12. Consult family physician at first onset of symptoms suggestive of influenza.

13. Spread this information as much as possible in newspapers, moving picture shows, school houses, churches, etc.

F. G. PERNOUD,
Medical Advisor Southwestern Division,
American Red Cross.

Service, it suggested the following, "The value of fresh air through open windows cannot be over-emphasized." F. G. Pernoud, medical advisor of the American Red Cross at that time, said, "Fresh air is always good. Keep your bedroom windows wide open, and secure as much sleep as possible."

As a result, the radiators were sized large enough to heat the room with the windows wide open. In fact, engineering manuals from the 1920s required designers to size the radiators to heat the room "on the coldest day of the year, with the wind blowing, and the windows wide open," according to Dan Holohan, heating expert of Heatinghelp.com. Can you imagine someone recommending that today? If you are replacing the heating system in a older building, the original designer may have sized the radiators similarly. When replacing the boiler, check the sizing of the existing radiators. If they are indeed oversized, you may be able to heat the room using much lower temperatures. In an old home close to me, we are able to heat it using 140 degree water in the cast iron radiators at the outdoor design temperature, with his windows closed, of course.

A hydronic heating system will surrender its heat to the building by many methods but the most common heat emitters are cast iron radiators, baseboard radiation, coils or radiant tubing. The following are some things to consider when re-using the existing terminal equipment.

When reusing the old radiators check to see if they are enclosed. Many older buildings or homes enclosed the radiators for either safety concerns or aesthetics. If the radiators are enclosed, the heat output is affected by as much as 30%. The chart to the right shows the effect of radiator enclosures on the heating capacity of the radiator.

Baseboard Radiation When looking at the baseboard radiation, visually inspect to see whether the internal damper is open or closed.

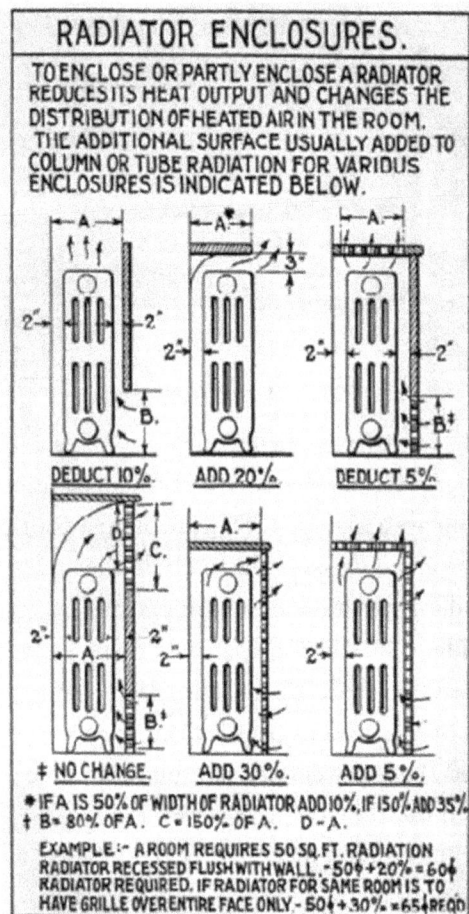

RADIATOR ENCLOSURES.

TO ENCLOSE OR PARTLY ENCLOSE A RADIATOR REDUCES ITS HEAT OUTPUT AND CHANGES THE DISTRIBUTION OF HEATED AIR IN THE ROOM. THE ADDITIONAL SURFACE USUALLY ADDED TO COLUMN OR TUBE RADIATION FOR VARIOUS ENCLOSURES IS INDICATED BELOW.

DEDUCT 10%.   ADD 20%.   DEDUCT 5%.

‡ NO CHANGE.   ADD 30%.   ADD 5%.

*IF A IS 50% OF WIDTH OF RADIATOR ADD 10%, IF 150% ADD 35%.
† B= 80% OF A.   C= 150% OF A.   D=A.

EXAMPLE:- A ROOM REQUIRES 50 SQ. FT. RADIATION RADIATOR RECESSED FLUSH WITH WALL.-50+20%=60♭ RADIATOR REQUIRED. IF RADIATOR FOR SAME ROOM IS TO HAVE GRILLE OVER ENTIRE FACE ONLY- 50+30% =65♭ RFQD.

Courtesy Dan Holohan

Most baseboard units have a manual damper that controls the convection through the internal coil.

Combining cast iron radiators and baseboard radiation is usually a control nightmare that results in comfort complaints. The cast iron radiator is slow to heat and slow to cool. The cast iron radiator will give off heat long after the burner or pump has shut off. The copper baseboard is fast to heat and fast to cool. The baseboard unit will heat more quickly on a call for heat than the cast iron radiator. When the burner or pump shuts off, the system heating is almost stopped. When these are combined, it make it very difficult to control. Think of it as Irreconcilable Differences. It would be better if they were on separate loops with their own pump and thermostat. If the thermostat is in the room with the copper baseboard, it may become satisfied and shut off before the cast iron radiator is even warm. Conversely, if the thermostat is in the room with the cast iron radiator, the room with the baseboard may experience wide temperature swings, resulting in comfort complaints.

Baseboard Radiation with Damper

Interesting Facts about cast iron radiators

- Radiators painted with metallic paint will give off 20% less heat than one painted with standard paint.
- If a radiator is painted with metallic paint, you could increase the output by painting over it with non metallic paint.
- Radiator heating output is about 2/3 convection and one third radiant heat.

Sizing Cast Iron Radiators

Most hydronic radiators were sized for 150 BTUH per square foot of EDR. The following chart shows the estimated BTUH per square foot per section. To find the BTUH for the entire radiator, multiply the BTUH per section by number of sections. This will give you the BTUH for the radiator based on 180 degree supply water. You could then compare the BTU rating versus the heat loss for the room to see if they are sized correctly. Do not assume that the radiators are all oversized. In a school that we service, someone had replaced a leaking cast iron radiator with one that was too small for the space. We had to add extra heat for the space to be comfortable.

Radiator Height	Number of Tubes						Number of Columns			
	3	4	5	6	7		1	2	3	4
13"					447					
16 1/2"					595					
18"									383	510
20"	298	383	452	510	723		255	340		
22"									510	680
23"	340	425	510	595			282	396		
26"	359	468	595	680			340	452	638	850
30"	510									
32"		595	735	850			425	566	765	1,105
36"	595									
37"		701	850	1,020						
38"							510	680	850	1,360
45"								850	1,020	1,700

Estimated Btu's per section of hydronic radiator
Based on 180 degree supply water & 70 degree F room temperature.

A cubic foot of natural gas burned will produce 8 cubic feet of nitrogen, 2 cubic feet of water vapor and 1 cubic foot of nitrogen.

Radiator Size Chart

1 Column	2 Column	3 Column

4 Column	5 Column	6 Column

Cast Iron 3 Column Radiator

Tube Type Radiator

Cast Iron Hydronic Radiator Ratings
Column Radiators

Column Radiators One Column		
Height	Sq Ft / Section	Btuh per Section
20"	1.5	255
23"	1.66	282
26'	2	340
32'	2.5	425
38"	3	510

Column Radiators Two Column		
Height	Sq Ft / Section	Btuh per Section
20"	2	340
23"	2.33	396
26'	2.66	452
32'	3.33	566
38"	4	680
45"	5	850

Column Radiators Three Column		
Height	Sq Ft / Section	Btuh per Section
18"	2.25	383
22"	3	510
26'	3.75	638
32'	4.5	765
38"	5	850
45"	6	1,020

Column Radiators Four Column		
Height	Sq Ft / Section	Btuh per Section
18"	3	510
22"	4	680
26'	5	850
32'	6.5	1,105
38"	8	1,360
45"	10	1,700

Cast Iron Hydronic Radiator Ratings
Thin Tube Radiators

Three Tube		
Height "	Sq. Ft. per Section	Btuh per Section
20"	1.75	298
23"	2	340
26"	2.11	359
30"	3	510
36"	3.5	595

Four Tube		
Height "	Sq. Ft. per Section	Btuh per Section
20"	2.25	383
23"	2.5	425
26"	2.75	468
32"	3.5	595
37"	4.125	701

Five Tube		
Height "	Sq. Ft. per Section	Btuh per Section
20"	2.66	452
23"	3	510
26"	3.5	595
32"	4.33	735
37"	5	850

Six Tube		
Height "	Sq. Ft. per Section	Btuh per Section
20"	3	510
23"	3.5	595
26"	4	680
32"	5	850
37"	6	1,020

Seven Tube		
Height "	Sq. Ft. per Section	Btuh per Section
13"	2.625	447
16 1/2"	3.5	595
20"	4.25	723

Connecting to the Combustion Air

Every fuel burning piece of equipment requires combustion air. When burning natural gas, a burner requires 10 parts of air for each part of gas for perfect efficiency. In real life, that is not realistic. We will add some air to assure safe operation. This is referred to as excess air. In industrial or large commercial boilers , we try for 15% excess air or about 11 1/2 parts of air for each part of gas. 10 plus 15% or 1 1/2 parts air extra. In smaller commercial boilers, it is usually more than that because it is difficult to control the fuel to air ratio with old style linkages. Consideration should be given for the conditions where the combustion

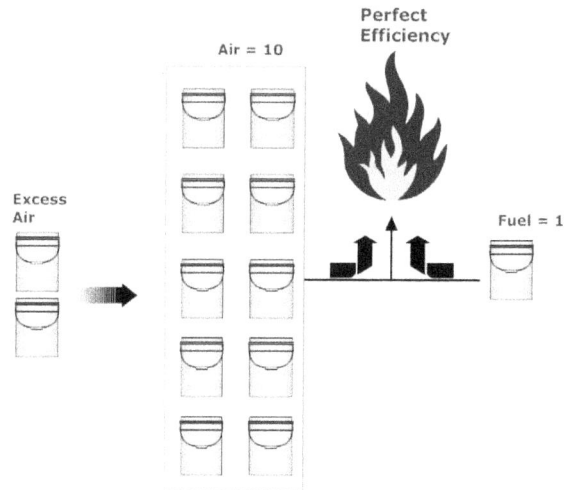

air is sourced from. At a factory we serviced, the combustion air was taken from same room as the process that was being performed which made it very dusty. That dust was drawn into the burner and affected the fuel to air ratio of the burner. We had to install a filter on the combustion air makeup to the burner. This increased the maintenance of the system. As the filter got dirty, it also affected the fuel to air ratio of the burner.

The older boiler rooms used openings in the outside wall to provide ventilation air. The International Mechanical Code suggests two openings in the wall for combustion. The first should be within 12" of the ceiling and the other within 12" of the floor. The following are the suggested

sizing for combustion air. I like using combustion air from the boiler room as my preferred method as the air density and temperature does not vary much.

The following are the combustion air sizing as per International Mechanical Code is as follows:

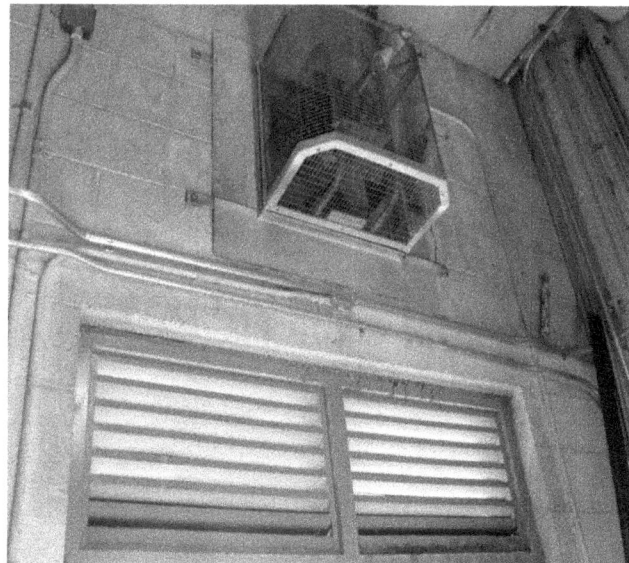

Size of Direct Openings
- 1" Free area for each 4,000 Btuh
- Horizontal Openings -1" Free space for each 2,000 Btuh

- Vertical Openings - 1" Free space for each 4,000 Btuh

- Mechanical Ventilation 1 cfm per 2,400 Btuh

Free Area When measuring the size of the existing combustion air louvers, you need to calculate the free area or the AK factor or free area of the louvers. A good rule of thumb is that metal louvers will have about 75% free area and wooden louvers will have about 50% free area. Let us assume that we have an opening of 20" x 20" or 400" of area. If the louvers are metal, the free area would be about 300". If the louvers are wooden, the free area would be around 200".

Make sure the grills are not plugged or covered. It is common to see cardboard or wood covering the combustion air louvers because it was too cold in the room. Check the outside of the opening as well because it could be plugged with leaves, grass clippings or garbage.

Verify that water pipes are not in front of the combustion air louvers.

Automatic dampers When using automatic dampers for combustion air, an end switch should be used to verify that they are open before the boiler is started.

Combustion air fan When using a combustion air fan, it should be sized for 1 cubic foot per 2,400 Btuh. For example, if all the fuel fired equipment inside the boiler room equals 1,000,000 Btuh, you would need about 417 CFM air. If using a combustion air fan, a control should be added to verify operation prior to the start of the boilers. In most instances, a differential pressure is which is used. The following chart may help you when sizing the combustion air fan.

Combustion Air Fan Sizing			
BTUH	CFM	**BTUH**	CFM
400,000	167	1,300,000	542
500,000	208	1,400,000	583
600,000	250	1,500,000	625
700,000	292	1,600,000	667
800,000	333	1,700,000	708
900,000	375	1,800,000	750
1,000,000	417	1,900,000	792
1,100,000	458	2,000,000	833
1,200,000	500	2,100,000	875

Direct Venting the Combustion Air

Should we use direct vented combustion air or boiler room air?

Direct vented is thought to be safer as you will not pull the room into a negative condition but it requires more maintenance. Let us assume that we perform our combustion testing in the fall using 80 degree weather and set the excess air for 15%. When the outdoor air temperature drops to 20^0F, the excess air percentage doubles, reducing burner efficiency.

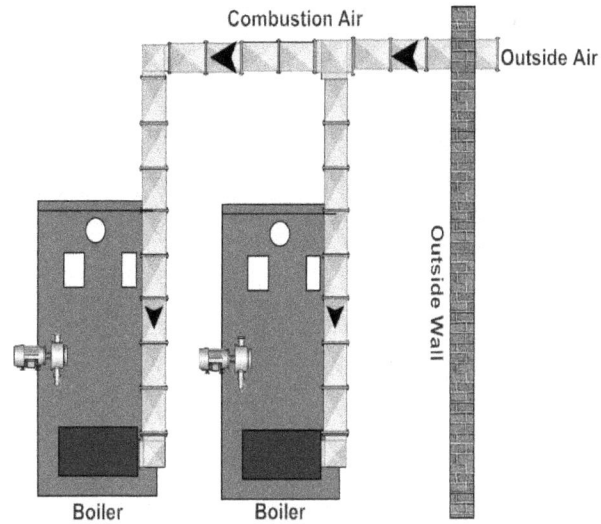

Combustion Air Temperature	Excess Air %
0	36.1%
20	30.8%
40	25.5%
60	20.2%
80 Burner Set up	**15%**
100	9.6%

When the outdoor temperature drops, the burner is less efficient due to the density of the colder air. I have been taught that the best time to adjust the fuel to air ratio for a burner is when it is cold. To properly adjust the fuel to air ratio, the flame has to stabilize. This usually takes 15 minutes. This is easier when the building has a high heat load and the burner can run longer. Now let us assume that we adjust the fuel to air ratio on the coldest temperature or 0 degrees F.

Combustion Air Temperature	Excess Air %
0 Burner set up	**15%**
20	9.7%
40	4.4%
60	-0.9%
80	-6.2%
100	-11.5%

From 40 degrees and up, our burner is operating in a dangerous condition. The carbon monoxide levels are probably excessive at that level and the boiler is in danger of sooting. To reduce the possibility of that happening, you should consider testing the fuel to air ratio twice per heating season if using direct vented combustion air.

77

When using ducted combustion air from the outside, Power Flame recommends sizing the duct for a pressure drop of 0.1" w.c. including all screens, filters, and fittings. Check with the boiler manufacturer for the suggested pressure drop that they want for their equipment. There is a chart in the rear of the book that will show you the estimated combustion air CFM for various boiler sizes.

Verify with the burner manufacturer about the material that is permitted to be used for the combustion air. Most installers use standard snap type duct for the combustion air. If you do, it should be insulated to avoid sweating in cold weather.

Insulated Combustion Air Duct

Connecting to the Controls

Oh, How Times Have Changed. When I was a young apprentice for a major control company, I was assigned to work with a journeyman that was a practical joker. He would send me for left handed pipe wrenches, pipe stretcher, water hammer, and of course, I had to get his coffee each day, black with two sugars. We had contracts to calibrate the pneumatic thermostats in many of the local schools. Pneumatic controls used compressed air for the controls. When you calibrated the old brass pneumatic thermostats, they had a small screw you would remove to insert the pressure gauge. From the time you removed the screw to the time you installed the pressure gauge, the air would hiss from the leak port of the thermostat. This was mirrored when you removed the gauge and reinstalled the screw. Everyone in the classroom would inevitably turn to see what was making the hissing sound. My journeyman would always use this opportunity to start coughing and gagging and whisper loudly, "The gas! the gas! I must stop it." The teacher would terrifyingly ask, "Is that gas? Oh my God!" The journeyman would tell the teacher not to worry and he had it under control, which would worry the teacher more. The boss would always yell at my journeyman after receiving a call from the irate school superintendant asking why we would release gas into a classroom full of students. He was relieved of his duties when a thermostat fell off the wall and the teacher evacuated the classroom and called the fire department thinking it was gas, which it technically is. As an ex control geek, I like working with the control techs in our industry. They have such a passion for their craft. If left alone, they will control or monitor everything that moves, slides, shakes, or switches. I like to reign them in a bit when controlling a hydronic heating system. There are several hard limits and a couple danger zones that should be heeded. Of course, these should be verified with the boiler manufacturer.

Design temperature - Most hydronic systems were engineered to supply 180 degrees F water at the outdoor heating design temperature.

Condensing temperature - Operating a standard non condensing boiler below 140 degrees F will allow the flue gases to start to condense, which could destroy the boiler, flue and chimney. It might also void the warranty. Another effect is that soot may build on the fire side of the boiler, causing a very dangerous condition.

Delta T - This is the temperature rise across a boiler. Most boilers are designed for a 20-25 degree temperature rise. If the system return water temperature is 150 degrees F, the supply water temperature to the system from the boiler should be 20-25 degrees F higher. A temperature span greater than the boiler design temperature rise could cause Thermal Shock inside the boiler.

Reset Ratio - The change in water temperature to the change in outside air temperature.

Outdoor Heating Design Temperature - This is the design temperature that engineers use when sizing a heating system. It used to be the coldest temperature that the locale experienced. It is now the temperature that occurs 2% of the time. This means that in a typical winter, the outside temperature will be at or cooler than the design temperature 2% of the time or about 175 hours per year.

Thermostat The easiest way to control a boiler is to connect it to a thermostat. While this may not be very sophisticated, it works on small buildings.

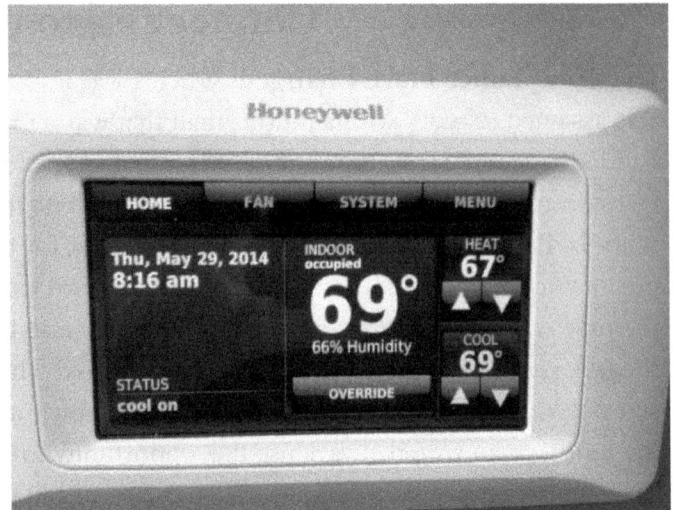

When using a night setback control, the recovery time is longer for radiators. In severe weather, you may have to start the occupied cycle several hours before the actual occupied time to allow the radiators to heat the space. On bitter cold days, consider eliminating the night setback all together. When using a thermostat, there is no outdoor reset of the water temperature.

Should we even use a night setback control? I found this interesting quote by York Shipley that may shed some light. "Day Night set-back operating systems amplify the potential for boiler damage from low temperatures and will use 25-50% additional fuel than a conventional primary secondary loop system working around the clock."

Warm Weather Shutoff This control uses an outdoor air sensor that will shut off the heating system and pumps when the outside air temperature reaches a certain degree, typically 55-60 degrees F.

Loosen control parameters To tweak more savings from a system, try loosening the control parameters. Unless you have a process that requires a tight temperature tolerance, allowing the temperatures to drift a bit will increase seasonal efficiency.

Reusing the existing controls. If you are replacing the boiler, I would urge you to consider changing the old controls to something that is more modern. Although the old pneumatic controls like the ones in the picture worked well, they have their limitations.

Building with univents Many schools used unit ventilators or univents for the heating and ventilation needs of the classrooms. As the name implies, they are primarily a ventilator and a heater second. If the building has univents, water flow is required all the time to each unit as the coil could freeze because they are subjected to the cold outside air. The univent is like a fan coil unit with an outside air damper. When we are controlling a hydronic boiler, these temperature limits must be followed. For example, when a reset control is installed on the boiler and the supply temperature is below 160 degrees F, the return temperature will typically be 20 degrees lower or 140 degrees F. At that temperature, this allows the flue gases to start to condense. Extended operation below that temperature could cause damage to the boiler. Many standard efficiency boiler manufacturers will void their warranty if operated below 140 degree F.

Thermal purging - On small hydronic systems with large mass boilers, this has been a very cost effective solution. We will use either a two stage thermostat or a time delay relay. On a call for heat, the system circulator starts. If after a certain amount of time the thermostat is still calling for heat, the burner starts. This allows the system to use the residual heat in the boiler to heat the building. If there is not enough heat, the burner will start.

Connect Circulator to Fan Terminal - Most commercial thermostats allow continual fan operation during the occupied time. On smaller commercial buildings, we will sometimes control the pump using the fan terminal of the thermostat. During the occupied setting on the thermostat, the circulator will operate continuously. During the unoccupied time, the pump will only operate when there is a call for heat.

Lead-Lag Controls are used to change which boiler is the lead boiler in a series of boilers to even the wear of the boilers. When using non condensing boilers piped in a primary secondary system, I do not like to switch the lead boilers continuously. If you do, you are always firing into a cold boiler, allowing the flue gases to condense. I like to switch the lead boilers monthly or yearly. If you use the same boiler throughout a month or season, you will always be firing into a warm boiler. It will prolong the life of the boilers. I prefer using this strategy for condensing boilers as well because I like to use the heat stored in the boiler for the loop rather than allowing it to be unused and heating another boiler full of water. On low mass boilers, you may be able to switch the lead more often. If you have multiple boilers in a long row, be careful when allowing the one furthest from the chimney to be the lead boiler as the flue gases may condense before reaching the chimney in light load conditions.

Reset Control As we discussed earlier, most hydronic systems were designed to heat the building using 180^0F at the outdoor design temperature. Someone smarter than me realized that as the outdoor temperature rises, the building could be heated with cooler water. That is how a reset control came to be designed. It adjusts the boiler temperature down as the weather warms. A typical reset control ratio is a One to One

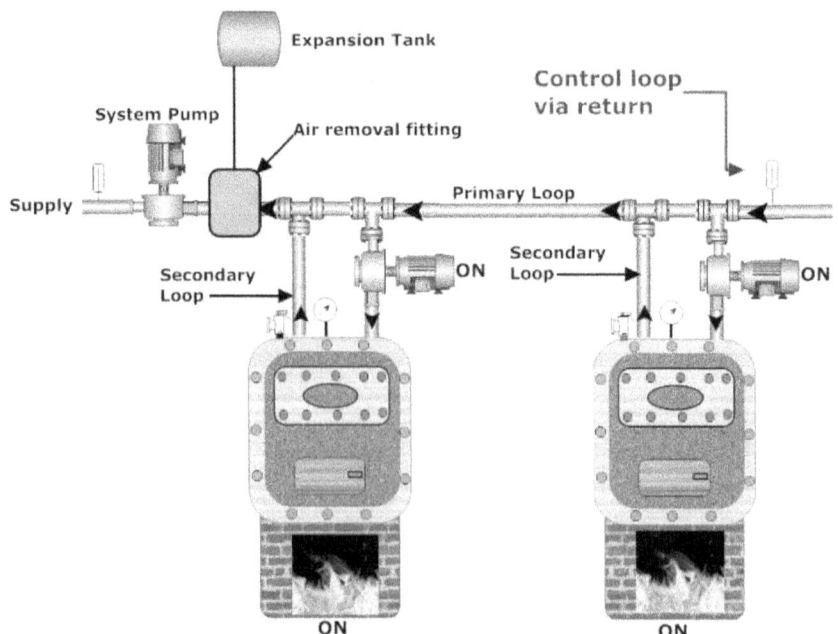

ratio. That means that for every degree warmer it is outside, the boiler water temperature is dropped a degree. I have found most newer boilers heat so quickly that the sensor that provides feedback to the reset control should be located on the return piping. If it is on the supply, the control is satisfied too quickly and causes short cycling of the boilers. When installed on the return, the control will not sense a temperature change until the water goes through the entire loop.

Typical One to One Reset Ratio		
OA Temperature	Hot Water Supply Temperature	Return Water Temperature
0	180	160
30	150*	130*
60	120*	100*

Standard boiler could be condensing at that temperature.

Bypass Valve A bypass valve is sometimes used to introduce warm supply water into the return loop to keep the boiler above the dew point or condensing temperature.

Hybrid system controls A hybrid heating system consists of a heating plant that mixes standard efficiency boilers with condensing boilers. The advantage to this type of system is that the owner gets the efficiency of a condensing boiler with the longevity of a standard efficiency boiler. According to the CIBSE or Chartered Institute of Building Service Engineers, the life expectancy of a condensing boiler is about 10-15 years. According to ASHRAE, the life expectancy of a standard boiler is between 24-35 years. When controlling this type of system, I like having the standard efficiency boiler(s) as the lead boiler when the water temperature is above 140 degrees F with the condensing boiler as the lag or backup boiler. When the reset control based on outdoor temperature and building heat loss calls for a water temperature below the 140 degree F threshold, I like having the condensing boiler as the lead boiler and the standard boiler as the lag boiler. In the picture, the boilers on the left are the standard efficiency ones and the one on the right are the high efficiency condensing boilers.

According to the 2009 International Energy Conservation Code 503.4.3.4 Part Load Controls: Hydronic Systems greater than or equal to 300,000 Btuh output shall

1 Reset water temperature by 25% of the design supply to return water temperature difference.

2 Reduce system water flow by 50%

This could be substantial savings considering most pumps operate 24 hours a day during the heating season. While this a good idea, care should be exercised when reducing the flow through boilers. Many boilers require a minimum flow to avoid damage to the boiler. A common control technique is to adjust the flow for a 20 -25 degree rise across the boiler. Consult with the boiler manufacturer when changing the flow to assure adequate GPM flow for the boiler.

Pump Speed Control	Pump Speed Control

Zone Valves An apartment building that we serviced had a zone valve for every apartment. While this seemed to make sense to the installer, it did not work well. When several of the zone valves were closed, the flow through the open valves was excessive resulting in noisy operation. It sounded like the call of orca whales. Yes, I watch Animal Planet. This high velocity would not allow the water to give up its heat. The boiler short cycled and the apartments that were cold stayed cold. To remedy the problem, we had to disconnect the control wiring, manually open all the electric zone valves, and balance the flow. We balanced the flow by measuring the temperature drop from the supply to the return. We adjusted the flow for a twenty degree drop. We then installed a reset control and a remote thermostat in one of the apartments. In addition to being unable to control the building properly, the closed valves could damage the pump by dead heading the pump. When the zone valves were not open, there was insufficient flow through the boiler, damaging that. If I was servicing this building today, I would have installed a variable speed control on the circulator that would slow down when only a zone or two called for heat.

Radiator valves are commonly installed on radiators. Some of them are the self contained ones that use thermal expansion to close the valve. The drawback to radiator valves is that it could cause the pump to dead head if they all shut off at the same time. It could also cause excessive flow through the open valves if the pump is operating at the same speed.

Dead Heading a pump is when all the flow is stopped for the pump by closing the valves. On the older pumps with positive displacement, this could have caused damage to the pump and mechanical seal. The caution with the smaller circulators today is that the water temperature could raise high enough from the pump impeller spinning to cause the water to

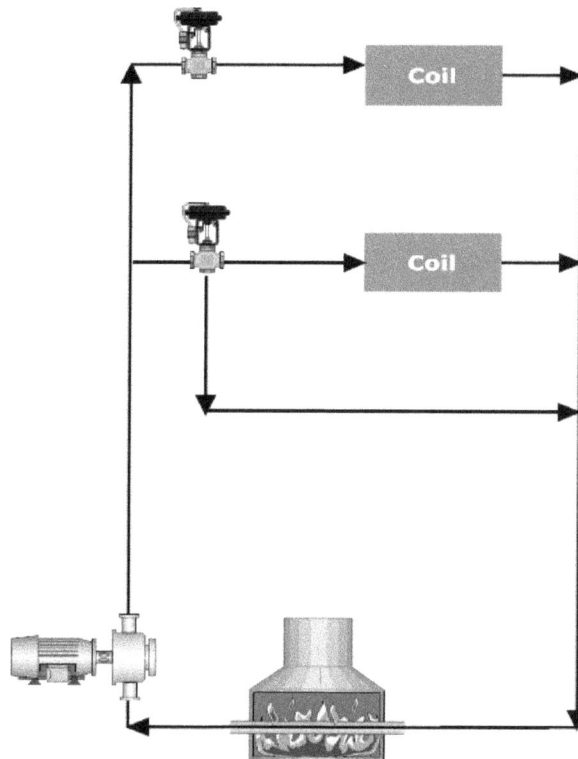

3 way valve with 2 way valves

flash to steam, ruining the pump.

When two way valves close, the system pressure may rise and could affect the pump. Pressure bypass is sometimes used when using two way valves. The bypass control senses the pressure in the piping and will open a bypass valve to allow the flow back to the pump inlet. This reduces the chance of excessive velocity through the piping. Another way to eliminate the excessive flow is to use three way valves instead of two way valves as a way to control the velocity.

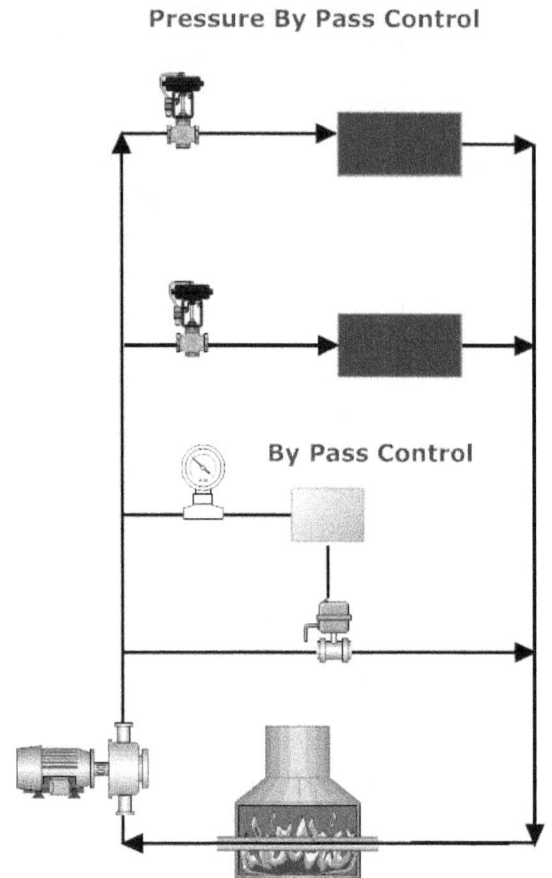

Pressure By Pass Control

By Pass Control

Three Way Valves Most boilers are designed for a twenty to twenty five degree rise through the boiler. In some instances, three way valves are problematic for hydronic systems. They could damage the boilers. There are two types of three way valves. The first is a mixing valve. This valve has two inlets and one outlet. It will mix the temperatures from the two inlets to the desired temperature outlet. The other type valve is a diverting valve. This valve has one inlet and two outlets. These are usually for bypassing an object such as a boiler in the summer or a chiller in the winter. They are typically a two position valve meaning that they are either open or closed.

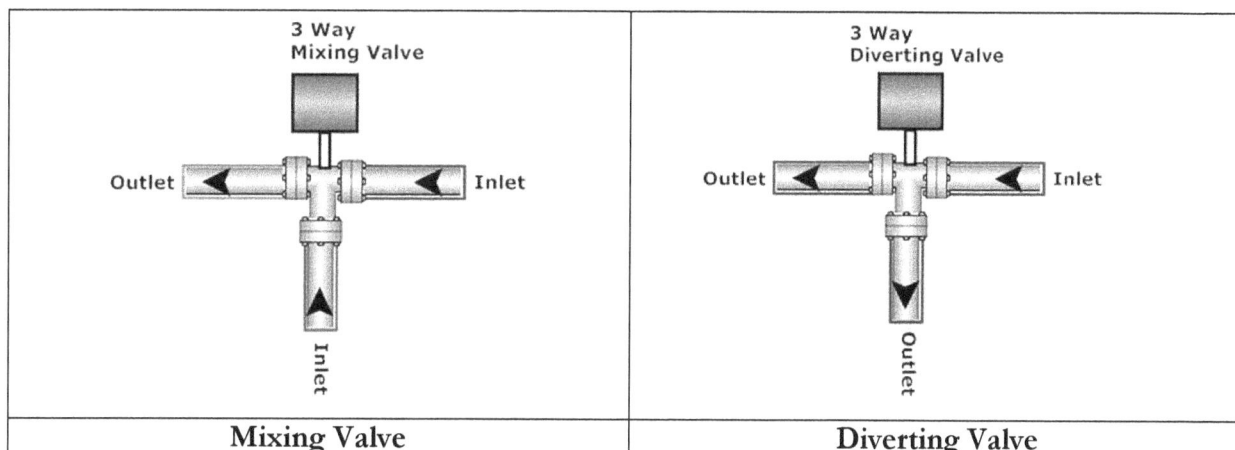

Mixing Valve	Diverting Valve

Bypass pump When using a three way valve in a hydronic heating system, thermal shock could occur. To limit the chances of thermal shock, some boiler manufacturers recommend using a bypass pump to warm the return water. The pump will take hot supply water and inject it into the return piping before the boiler. The water has to have enough room to mix properly to avoid thermal shock so some manufacturers want the connection to the return pipe to be 10-20 pipe diameters upstream of the boiler. If the boiler return connection is a 4" one, the connection must be 40-80" upstream. This could really be an expensive project as it may require complete re-piping of the boiler piping. Weil McLain has some suggestions when using a 3 way valve for their boilers. The GPM of the bypass pump should be 1/4 to 1/3 of the system pump GPM. If the system pump is rated for 100 GPM, the bypass pump should be between 25-33 GPM. The minimum timing of the 3 way valve from one extreme to the other should be four minutes. All circulators must run at the same time.

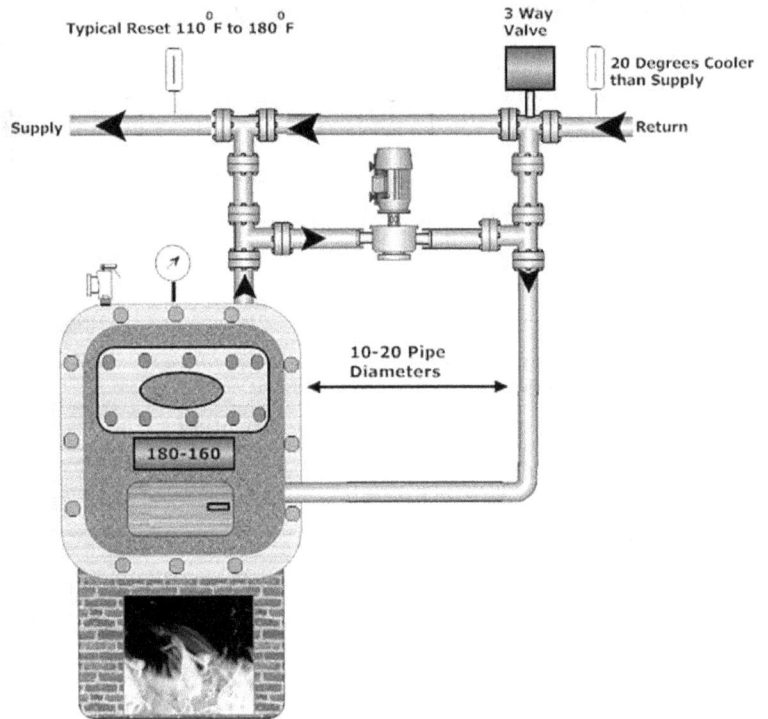

The picture below shows the effects of an improperly installed three way valve. Each year, the last section of the boiler cracks from thermal shock.

Connecting to the Gas Piping

Verify the gas pressure that is available for the building. If the existing boiler had an atmospheric burner, it may have used only a couple inches of gas pressure. The new burner may require up to 14" or 1/2 pound of gas pressure. On the other end, high gas pressure could also be a problem. On one project, the old boilers used high pressure gas over 1 pound of pressure. This was unknown at the time of ordering the new burners. The installer had to purchase and install expensive gas regulators at a cost of several thousand dollars. Verify that the pipe size is correct as well.

Many manufacturers do not want Teflon pipe dope or tape on the pipe joints for the gas fittings. Read the installation manual to see if it is allowed to be used on the new boiler. The following is an excerpt from the installation manual of a Webster burner regarding the use of Teflon:

"Warranties are nullified and liability rest solely with the installer when evidence of Teflon is found."

What about the existing Teflon? In my seminars, many installers will ask me what to do if the system has Teflon pipe tape on the existing fittings. Quite frankly, I believe that you should replace it with a non Teflon substance if the manufacturer will void the warranty if Teflon is found. If the manufacturer is adamant about the use of Teflon, I would not want to chance a dangerous condition so I would remove it. The other question I get asked is whether you could use the yellow Teflon tape that is advertised for use on gas lines. You really have to make that decision but I would not use that either if I was installing the above burner. There is not a small asterisk on the warranty sheet that says that you could use it. I like using Teflon as a pipe thread sealant personally as I think it provides a better seal. I would just not want to risk a catastrophe because I used a substance that the manufacturer said not to use.

Gas Pipe Size This can directly affect the way a boiler operates. If the piping is too small, the burner will not be able to fire at capacity. It could trip the gas pressure switch if the burner has one.

Gas Train The gas train is a series of components that consist of electric and manual gas valves, gas pressure regulator, and safety controls to assure safe firing of the boiler.

Dirt Leg

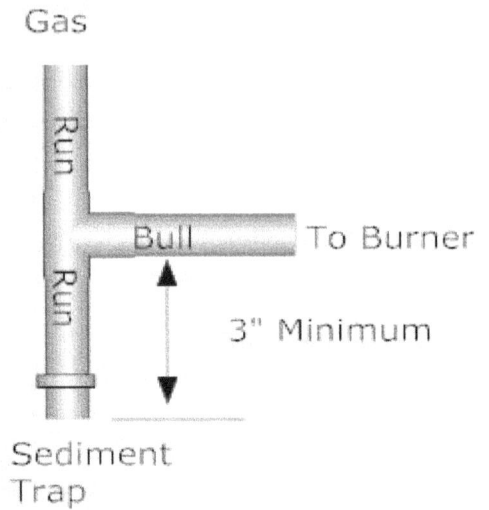

The first component of the gas train is the sediment trap or more commonly referred to as the **"dirt leg"** This is to catch any dirt or moisture that may be entrained in the natural gas feed to the building. It is typically a nipple and cap that is located on the run of the tee. The dirt leg should be at least 3" long. The feed to the gas train is on a right angle coming from the bull of the tee.

ASME CSD1* recommends a **strainer** before the other horizontal components in the gas train. Although not commonly installed, it is a good idea as it will also protect the gas train components from dirt. A medical facility in my area found the benefit of a strainer when dirt from the gas line lodged under the seat in the gas valve, filling the boiler with gas. When the burner ignited, the boiler exploded. The brand new boiler and burner had to be replaced. Luckily, no one was injured.

The upstream **manual gas valve** is just downstream of the strainer, if used. This manual gas valve usually has a pipe connection on the upstream side of the valve for the pilot connection. (see arrow) That pipe tapping is where you should attach the pipe or tubing to the burner pilot assembly. The pilot connection should be on the upstream side of the valve to allow you to test the safety controls. The valve handle should be oriented so that the handle is parallel to the gas flow when open and perpendicular to the gas flow when closed. It almost looks like the handle is pointing to next component if open. This allows you to quickly know whether the gas valve is open or closed.

Gas train vent sizing is covered on page 144.

The next item in the gas train progression is the **gas pressure regulator**. It should be upstream of all electric valves. The orientation of the regulator is important and is usually right side up with the adjustment access straight up. Inside the top of the regulator is an adjustment screw and a spring. On many regulators, the spring can be changed to allow different gas pressures to the burner. The gas pressure regulator may have a fitting on top that is to be vented to the outside. This vent pipe can sometimes plug and fill with spiders or bees. If plugged, it will not allow the gas to flow. A bug screen should be installed on the outdoor termination of the vent pipe. Some new regulators use vent limiters which do not require venting to the outside.

The flames would roll out of the side of the boiler when the burner started. It would actually roll so far out of the boiler that it would burn the boiler jacket. The cause of this anomaly was because the gas pressure regulator was improperly installed downstream of the first electric gas valve. When the electric gas valve was powered off, the gas pressure regulator would open wide. After the electric gas valve was energized and open, the regulator would be wide open and over feed the gas to the burner. The gas pressure would eventually regulate itself to the proper setting but not until the flame shot out of the side of the boiler. To repair the problem, the regulator had to be relocated upstream of the electric valves. This was shipped from the factory like this.

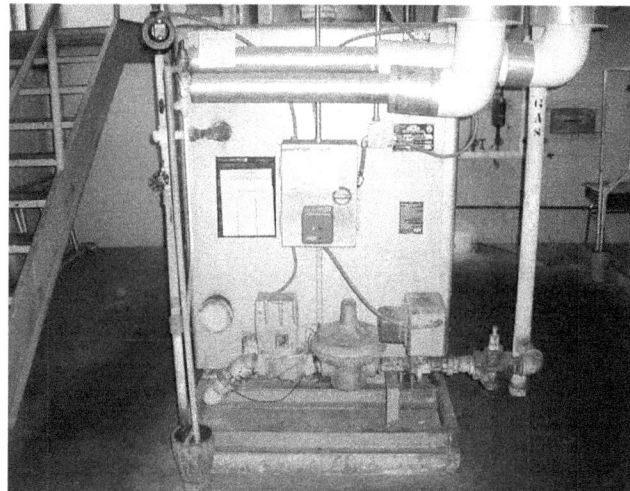

In some areas, ASME CSD1 recommends a relief valve in the gas train. I have never seen one on any of my projects.

The next item downstream in the gas train progression is the low **gas pressure switch**. As you can guess, this control will shut off the burner if the gas pressure is below the control setting. A high gas pressure switch will be located between last manual gas valve and the burner. Most are installed on the top of the burner before it goes into the boiler. This control will shut off power to the burner if the gas pressure is too high. These gas pressure switches typically have a manual reset that will require resetting if they trip. Gas pressure switches are not installed on every boiler and are typically used when the gas rating of the burner is greater than 2,500,000 Btuh.

Low and High Press Gas Cut-off

Electric safety shutoff valves are the next components the gas train. It is most likely two redundant valves. Redundant gas valves are used to limit the possibility of gas leaking through the valve and filling the boiler, creating a dangerous situation. If the state or municipality adheres to ASME CSD1, there will be a tapping with a valve and cap immediately after each valve. This is to allow the leak testing of the gas valves. Many facilities will leak test the gas valves using a bubble test. The orientation may be important when installing the gas valves and the installation manual should be consulted to see whether the valves can be installed horizontally or vertically.

Some older gas trains used what was called Block, Block and Bleed gas train. In between the two Normally Closed safety shutoff valves was a tee and the pipe from the bull of the tee was attached to a solenoid valve that was Normally Open. It was installed to vent any gas that may leak past the first electric safety shutoff valve to the outside. If the system had this type of gas train, the outlet pipe from the solenoid valve was to be vented by itself to the outside. The other gas train components could be combined into a single vent but separate of the Bleed valve piping. The solenoid valves should be checked on a regular basis to assure that gas is not leaking through the solenoid valve.

Adjustable Orifices are sometimes used to convert an On Off diaphragm gas valve to a slow opening valve. It is like a poor man's version of a low high off gas train. This slow opening helps to quiet the burner ignition and reduce the chances for rumbling in the stack. If you have a system that is rumbling, you may want to try using one of these. They are installed in the vent connection to the gas valve. The adjustment screw determines the opening speed.

"Our gas costs are suddenly very high. Do you have any suggestions?" The customer asked. I asked what had changed and was assured that nothing had changed. There was something wrong with my boiler that was over 10 years old I was informed. When we got there, we saw that something did change. Someone had replaced the Normally Open Bleed valve

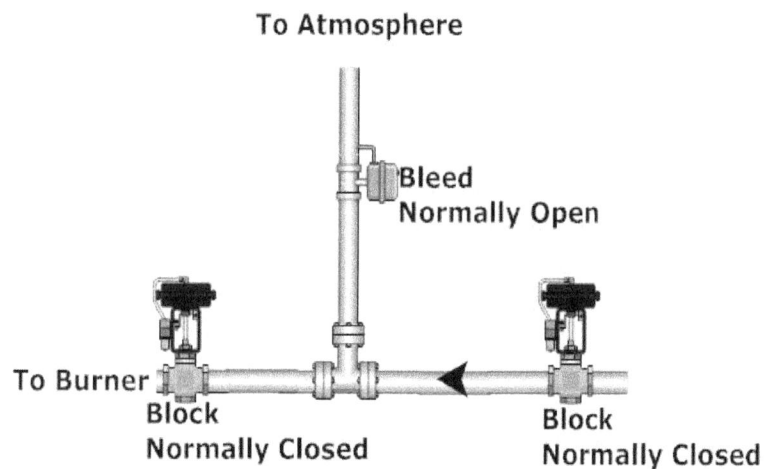

To Atmosphere

Bleed
Normally Open

To Burner

Block
Normally Closed

Block
Normally Closed

with a Normally Closed solenoid valve. No one knew who changed the valve. It must have been the Boiler Fairy that anonymously changes components and control settings. The solenoid valve would open when the Safety Shutoff valves opened, allowing a 1 1/4" pipe filled with gas to vent to the outside anytime the boiler was firing. Once the valve was changed back, the client liked my boiler again.

Another **manual gas valve** or the downstream manual gas valve is installed right

93

past the last electric valve and before the modulating or firing rate valve.

Firing rate valve. This valve is operated by a modulating motor and will change according to input from the modulating control. These are not installed on an on off burner.

Typical gas train

For perfect efficiency, a burner requires 10 parts of air for each part of natural gas.

Gas Valve Leak Testing

*ASME CSD1 - American Society of Mechanical Engineers Controls and Safety Devices for Automatically Fired Boilers requires that you should be able to leak test the gas valve to check for gas leakage through the valve. The following picture shows how one contractor piped the gas train to allow for leak testing.

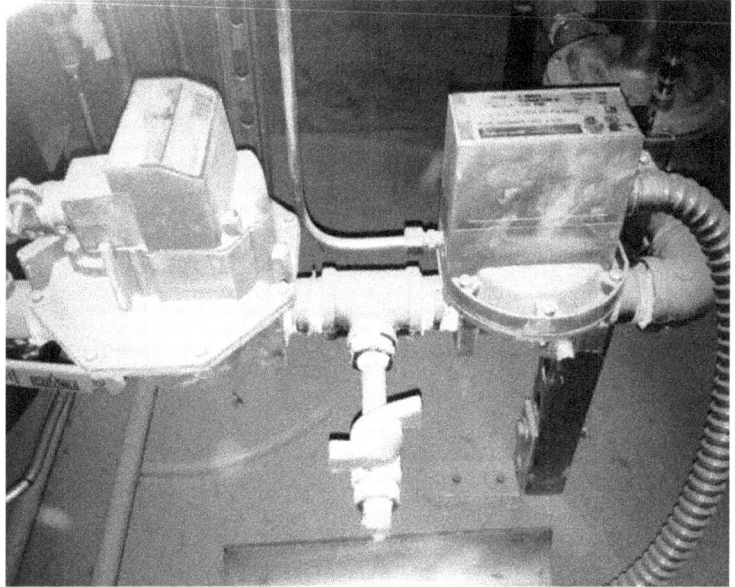

It amazes me that we would never tolerate a leaking water valve on the sink but a gas valve is permitted a certain

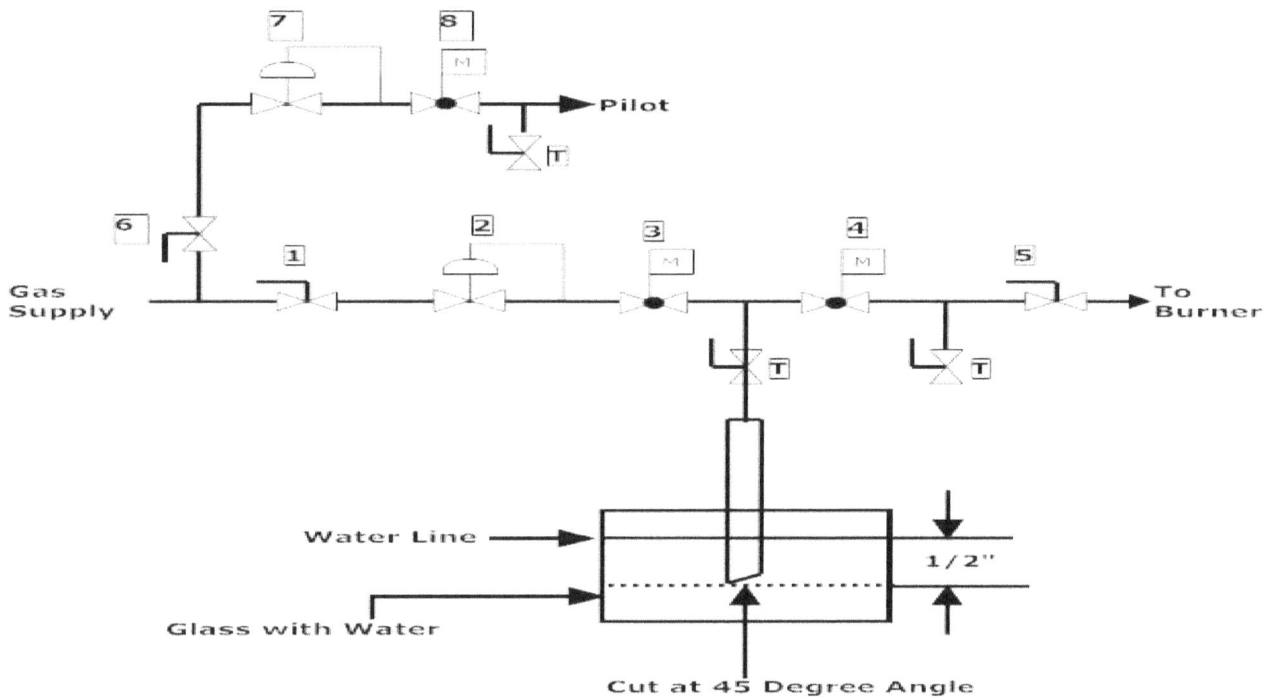

Water Line

Glass with Water

1/2"

Cut at 45 Degree Angle

1	Upstream Manual Gas Valve
2	Gas Pressure Regulator
3	Safety Shutoff Valve
4	Safety Shutoff Valve or Blocking Valve
5	Downstream Manual Gas Valve
6	Manual Gas Valve- Pilot
7	Pilot Gas Pressure Regulator
8	Pilot Safety Shutoff Valve
T	Test Ports These may be part of valve

amount of leakage. And how do we test for a leaking gas valve? We count bubbles. Really!

According to ASME CSD1, as well as most gas valve manufacturers, provisions should be made in the gas train to allow for leak testing of the electric safety shut off valves (SSOV). This is usually a ¼" tapping in the downstream nipple after the gas valve. CSD1 and the manufacturers recommend testing the valves at least once per year. If there are two electric SSOV's, then both have to be tested. The pilot solenoid valve should also be tested at least once per year. Some valves will have a downstream tapping as part of the valve. If the gas train contains a bleed valve, I would suggest testing this at the same time. This valve is piped to the outside and unless it is tested, a leak may never be found.

When testing the valves, refer to the manufacturer's recommendations. To meet the U.S. requirements, leakage must not exceed ANSI Z21.21, Section 2.4.2. It is based on air at standard conditions and limits leakage to a maximum of 235 cc/hr per inch of seal-off diameter. This is not the same as pipe diameter. The following is the maximum bubble count for the valve sizes.

Valve Size (Inches)	Allowable Leakage (cc/hr)	Maximum Bubbles per 10 second test.
¾"	458	16
1'	458	16
1 ¼"	458	16
1 ½"	458	16
2"	752	26
2 ½"	752	26
3"	752	26
4" & Larger	1,003	35

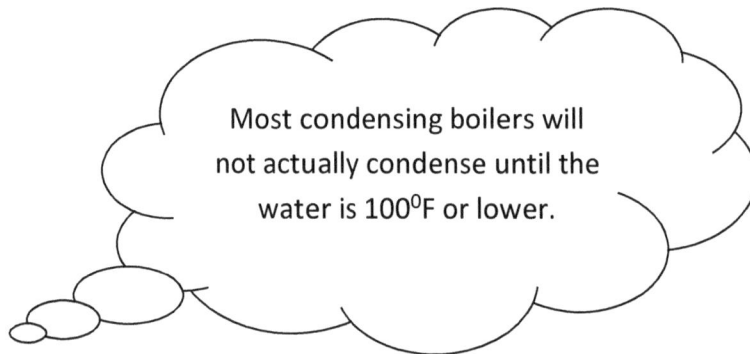

Most condensing boilers will not actually condense until the water is 100°F or lower.

Connecting to the Electrical

When replacing an old boiler with an atmospheric burner with a new boiler that has a power burner

and secondary pump, you will need to find more power. Verify that the existing power is adequate for the new burner. This includes voltage and electrical phase requirements. An electric sub panel may be required. Notate the electrical panel information to see if new breakers are available for the existing panel. In some instances, they may not be available. It is a good idea to take pictures of the existing electrical panels.

"Your boiler is not working" was the anxious call I received on the first cold day of the heating season. When the technician arrived, the boiler was working fine. He checked all the operating and safety controls. The boiler ran perfectly. He showed the customer and then left. A few hours later, that same customer called and said, "The boiler stopped working as soon as the technician left. Does he know what he is doing?" I bit my tongue and told the client that I would meet the technician there. I arrived before the technician and verified that the boiler was indeed running. We showed the client and he smirked and said "It works now but just wait until later. It knows you are here." As we were leaving, I shut off the light switch to the boiler room and heard the boiler shut off. It then dawned on me that my electrical meter was on top of the boiler. I switched the lights on and heard the boiler start. I switched the lights off again and the boiler stopped. I turned the lights on

and heard the boiler start and smiled. "Have you had any electrical work done here lately?" I asked the client. He replied that yes but assured me it had nothing to do with the boiler. I showed him that the light switch for the boiler controlled the boiler. We changed the wiring back and the boiler ran great after that. The client then asked if we were going to invoice him for the service call.

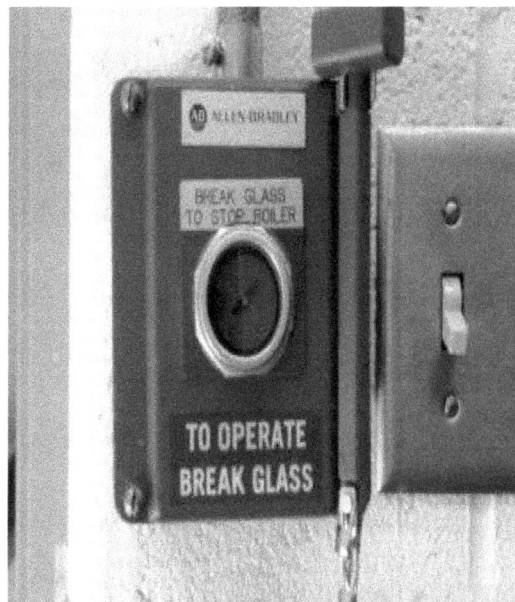

Door Switch American Society of Mechanical Engineers - Controls and Safety Devices or more commonly referred to as ASME CSD1 code calls for a switch to be installed at each exit from the boiler room. The switch is used to cut the power to the boilers inside the boiler room in the event of a malfunction. These are a great idea even if your municipality does not follow the code. If a boiler were to malfunction, it could fill the room with steam. In addition to causing near zero visibility, steam displaces oxygen so breathing is difficult.

The fire department received a call from an elementary school when the smoke sensor tripped. They arrived at the school and opened the boiler room door. They were greeted with what they thought was white, moist smoke. Upon hindsight, It was later thought to be steam leaking from the old boiler. The first responders used their machinery to cut the metal jacket from the boiler. They trained their fire hoses on the glowing cast iron boiler sections and opened the brass valves. This caused the cast iron sections to crack. The first responders were very lucky that this did not cause a catastrophic event. When steam is quickly cooled, a phenomenon called Condensate Induced Water Hammer occurs. This sudden condensing of the steam creates pressures in excess of 1,500 PSI. The boiler was rated for only 15 pounds of pressure and you could see how this could have been much worse.

Connecting to the Flue

In an effort to reduce the installation costs, the installer opted to reuse the existing stack. The existing stack was a Category III, which meant that it was a non condensing, positive chimney and flue. The new boiler called for a Category I which was non condensing with a negative chimney. After about two heating seasons, the Category III flue collapsed into itself because it was not designed to operate in a negative condition. It was like a beer can being squeezed after consumption. It was a legal mess and could have been a dangerous situation. The entire flue had to be replaced at the expense of the installer and designer. If you are re-using the existing flue, verify that the size and type is correct

Flue Categories

Category I Appliance operates with a negative vent static pressure and at temperatures above condensing temperatures. Typical efficiency range 78% - 83%.

Category II Appliance operates with a negative vent static pressure and at or below condensing temperatures.

Category III Appliance operates with a positive vent static pressure and at temperatures above condensing temperatures. Typical efficiency range 78% - 83%.

Category IV Appliance operates with a positive vent static pressure and at temperatures at or below condensing temperatures. Typical efficiency are over 90%.

Check the size of the flue that the old boilers were using. If the old boiler had a draft hood, the flue sizing may be oversized when using a power burner. I suggest making a small diagram of the existing flue and consult a specialist when reusing the existing flue.

There are a couple other things to check while you are on the job site. Are other items, such as a pool heater or domestic water heater, vented into the same flue? When you remove the venting from the flue and leave the existing water heater or pool heater in the old chimney, it is referred to as an orphan. You may have to change the venting of the old water heater or install a flue liner for it. Always remember the "7 Times Rule" which is: The flow area of the largest common vent or stack shall not exceed seven times the area of the smallest draft hood outlet. Since most water heaters use a 3" flue, the largest area to connect the water heater should be 49" in area or a 7" round one. The following is a chart that shows the largest common round flue that each can be connected to.

7 Times Rule Round Flue			
Smallest Draft Hood Outlet	Largest Common Flue	Smallest Draft Hood Outlet	Largest Common Flue
3"	7"	9"	22"
4"	10"	10"	26"
5"	12"	12"	30"
6"	14"	14"	36"
7"	18"	16"	42"
8'	22"		

If you are connecting to a square or rectangular chimney, you will need to estimate the areas. The following is a chart that shows the sizes that the vent can be connected to:

7 Times Rule Rectangular Flue		
Smallest Draft Hood Outlet	Area of Vent	Largest Common Vent Area
3"	7.06"	49"
4"	12.56"	88"
5"	19.64"	137"
6"	28.27"	198"
7"	38.48"	269"
8"	50.27"	352"
10"	78.54"	550"
12"	113.1"	792"
14"	153.94	1,078"
16"	201.06"	1,407"
18"	254.46"	1,781"
20"	314.16"	2,199"

Allow me to show you an example: If you remove the boilers from a chimney that is 12" x 12" square and want to see if we can leave an old water heater with a 4" flue outlet. The 12" x 12" chimney is 144" This is greater than the rule of 7 sizing for the 4" flue which is 87.92". On this job, we would have to make provisions for the water heater flue. A chimney liner would most likely have to be installed here. You better have that price in your quote.

Horizontal vs. Vertical Another rule to remember is that the horizontal vent must be no more than 75% of the vertical height of the flue. If using B Vent, the horizontal length can be the same length as the height of the chimney, as per International Fuel Gas Code, 2006 503.10.9

Effect of Draft on boiler efficiency Draft occurs when the chimney is heated. The warmer air is lighter and rises. The cooler air in the boiler room replaces the warmer air and this flow of air is called draft. The draft is the measurement of the velocity of the flue gases through the boiler and chimney. If the velocity is too slow, you will have rollout of the flue gases. If it is excessive, it could actually pull the flames into the boiler or pull the heat from the boiler. We were called to a housing authority because the water level was bouncing inside a low pressure steam boiler. We checked all the usual causes for this such as burner over-firing, feed water capacity, water quality, and boiler draft. Our draft reading was -0.5" w.c. and our boiler was designed for -0.05" w.c. The draft was ten times the level suggested by the boiler manufacturer. This excessive draft caused the flue gas velocity to be too high and affected the boiler efficiency as the flue gas did not give up its heat into the boiler, as evidenced by the elevated stack temperatures. We also saw that the excessive draft actually pulled the flames into the rear of the boiler. This caused the rear of the boiler to be hotter than the front, resulting in large waves of water from the rear crashing to the front of the boiler. It would trip the low water cutoff. Once we got the draft back to the recommended setting, the boiler water settled down and operated as it was designed. To solve the problem, we suggested a larger barometric damper to reduce the draft for the boiler.

What is a Barometric Damper? One of the most misunderstood components inside a boiler room is the barometric damper. The barometric damper is used to control the draft inside a boiler. It is installed on boilers that use a Category I vent. To understand how the barometric damper operates, we need to understand what draft is. We all know that when air is heated, it will rise. When the hot air from the boiler rises up the stack, it must be replaced by the cooler air that surrounds the boiler. This cooler air will push the warmer air up the stack, causing flow. This flow of air from the boiler up the stack to the outside is referred to as draft. It is also called "Chimney Effect." The speed or velocity of the flue gas draft is affected by many conditions such as temperature difference between the inside and the outside of the building, wind fluctuations, chimney height, burner firing rate, and barometric conditions. It changes constantly. On cold days, the draft may be very high. In some instances, excessive draft could actually pull the flame off the burner. In lesser conditions, it could simply pull the heat more quickly than desired through the boiler. This wastes money as the heat does not have time to transfer to the boiler. Instead it is wasted up the stack. During cold startups, the cold stack may allow spillage of the flue gases into the boiler room until that stack warms enough to sustain draft.

Chimney Effect is sometimes experienced in tall buildings. Here, in my home town, a large commercial building used to have some real issues from the excessive chimney effect in the building. The main entrance to the building used to have a set of double doors with a small vestibule which was directly in front of the main escalators for the building. During the rush hour on a cold morning, the negative conditions inside the building would be very high. When both sets of doors were opened simultaneously, the wind would whip into the building at high velocities. These powerful gusts of wind would blow up the women's skirts and dresses, embarrassing and then angering the wearer. It was reported that the male observers would often display feigned outrage for the victims of the wind gusts. The building owners were forced to replace the double doors with large revolving doors in an effort to combat the building's chimney effect. On another building, the engineer factored the building draft into his HVAC design and made a system that used the chimney effect to provide free ventilation to the building without fans.

When a tall chimney is attached to a boiler, the draft readings will vary greatly. The boiler requires a stable environment and the chimney is like the "wild child", a true Odd Couple. Most older boilers were designed to have a draft at the outlet of the boiler to be about -0.05" w.c. The draft conditions inside the chimney could cause swings of ten, twenty or even one hundred times that amount. How do you provide a buffer to handle these erratic swings? A barometric damper is a great solution. High draft will pull the flue gases too quickly through a boiler, not allowing the heat to be transferred into the boiler. The barometric damper will be installed in the flue between the boiler and the chimney. It will be set for the desired draft conditions using weights and adjustment screws. If the draft inside the chimney is greater than the set point, the damper will open and allow air from the boiler room inside the chimney, rather than stealing heat from the boiler.

Draft controls are typically used when the stack or chimney height is greater than 30 feet. Excessive draft inside a boiler can cause other strange behaviors. In addition to increased operating costs, the high draft can cause flame impingement on the boiler. This could develop higher than desired levels of carbon monoxide. Flame impingement could also cause embrittlement of the boiler metal, lowering the life of the boiler. Embrittlement is the loss of ductility of the metal. It is like what happens when you keep bending a paper clip. It will eventually break. Flame impingement means that the burner flame is touching the metal surfaces of the boiler that were not designed to have flame directly on them. We had a vertical fire tube steam boiler where the flame was drawn into the rear tubes of the boiler. It caused violent surges and waves inside the boiler. This caused wet steam and tripping of the low water cutoff. Excessive draft can also cause the burner to over fire. If the gas pressure regulator is set for a certain pressure, the high draft can actually pull more gas through the regulator, over firing the boiler. Barometric dampers will only be installed on boilers with negative venting. Boilers with pressurized vents would spill flue gases out the barometric dampers into the room.

When choosing a barometric damper, there are two types; single and double acting. A single acting damper will have a stop that only allows the damper to swing one way. A double acting one will allow the damper to swing two ways. The single acting damper will close if there is a pressure inside the stack. The double acting damper will actually allow spillage of the flue gases into the boiler room in the event of blocked flues or down drafts. I know that sounds crazy. That is why I like seeing spill switches on the barometric damper. If there is spillage from the barometric damper into the boiler room, this switch will sense it and shut off the burner. In some locations, installation of spill switches is part of the boiler code.

Each fuel or combination of fuels requires a specific type of barometric damper. Single acting is traditionally used for oil fired burners and double acting is used for gas burners. The stops that are in the double acting dampers should be removed if only firing with gas.

How tall is your chimney? If you have ever been to a boiler room where they once had a coal fired boiler, the chimney is really tall. I have a theory that the reason these chimneys were so tall was to spread the coal ashes over a wider area. These stacks take forever to heat high enough to sustain draft. Once it is warm, the draw is tremendous. If you are connecting your new boiler to this dinosaur, there could be some real issues. The boiler flue gases could condense until the stack gets warm, causing rollout. Once the stack was warm, the draft could be very high. A good rule of thumb is that a draft control should be used if the chimney is over 30 feet high. The draft control may range from a barometric damper to sequencing draft control. It is not uncommon for Category I boilers to have flue gas spillage on the first one or two minutes when firing into a cold stack or chimney. A spill switch should be installed in the event of flue blockage.

Condensing Boiler Flues Most condensing boilers require a Category IV positive flue. This flue is a positive flue with condensing boilers and since it is positive, they cannot be combined with other flues like the older boilers did. Each boiler will require a separate flue, which increases the installation costs. In addition, the side wall of the building looks like an old pirate ship that had cannons every few feet.

When venting these boilers, exercise caution when using a common breeching. Since each boiler features a positive pressure in the flue, this could force the flue gases out of the idle boiler and damage the boiler. In addition, the flue gases could flow out of the idle boilers and into the boiler room. I have seen them combined when used with a chimney top exhaust fan to make sure the flue is negative.

"My bushes are ruined", the wife of the CEO of a large medical facility said angrily to her husband. The cause of her angst was that the new condensing boilers were vented directly above the bushes she had planted. When the flue gases condense, acids are formed when mixed with the water vapor. The flue gases, containing acids, landed on the bushes that the CEO's wife had purchased and planted. The bushes were destroyed and the CEO was furious as you can imagine. They wanted to get rid of the boilers and replace them with standard boilers that vented through the chimney. Be careful where you vent the condensing boilers. This cost the Director of Maintenance his job.

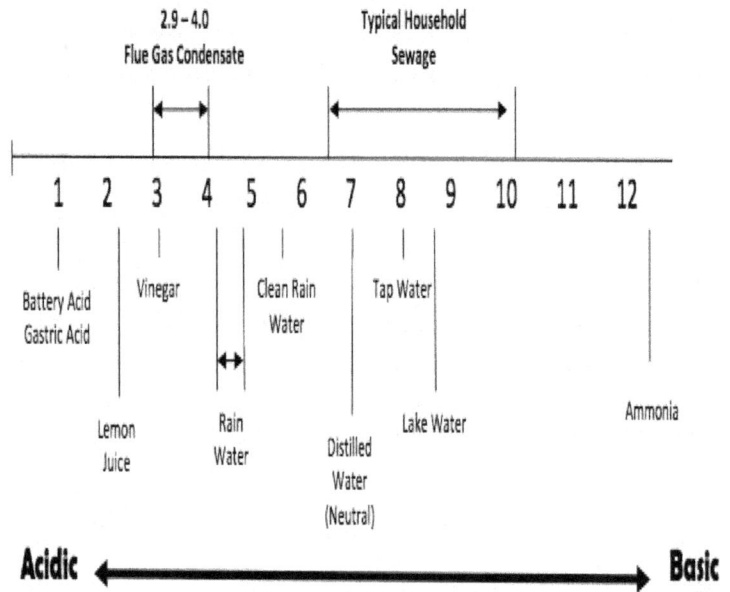

2.9 – 4.0
Flue Gas Condensate

Typical Household
Sewage

1   2   3   4   5   6   7   8   9   10   11   12

Battery Acid
Gastric Acid

Vinegar

Clean Rain
Water

Tap Water

Lemon
Juice

Rain
Water

Distilled
Water
(Neutral)

Lake Water

Ammonia

Acidic ⟷ **Basic**

Credit: Mike Bernasconi, Neutrasafe Corp.

Flue gas condensation When a boiler condenses, it means that the flue gas temperature drops low enough to fall below the dew point temperature. When this happens, the moisture that was part of the flue gas falls out and accumulates inside the boiler. The flue gases also have levels of sulfur and carbon dioxide, creating sulfuric and hydrochloric acids when they combine with the water vapor. Due to the acid formation, the pH lowers. It is typically between 2.9 to 4.0 A reading of 7 is neutral. While it appears to be an insignificant amount, pH readings are logarithmic, which means that a reading of 6 is ten times more acidic than a reading of 7. A reading of 5 is 100 times more acidic than a reading of 7. So you can see that a reading is 2.9-4.0 is vastly more acidic than the neutral reading of 7. Many ask how that would affect the piping. Consider this, when the pH level is greater than 8 or more which is alkaline, copper pipes form a copper oxide film, protecting them from corrosion. When the pH is lower than 8, the copper oxide film in the copper pipe dissolves and the pipe is vulnerable to corrosion, resulting in leaks.

The volume of the condensate is also another consideration. When in a fully condensing mode, a boiler will produce about one gallon of acidic water mix for each 100,000 Btuh burned. If you have a 1,000,000 Btuh boiler, it will produce about 10 gallons per hour. Many municipalities do not allow that acidic mixture into the drain as it could destroy the pipes. To avoid that, you may have to install a condensate neutralizer. These have to be inspected yearly. The following are produced when you burn one cubic foot of gas:

- One cubic foot of carbon dioxide
- Two Cubic feet of water vapor
- Eight cubic feet of nitrogen

Venting Materials

Many of the newer boilers will require a change in the venting of the boiler. The new boilers may require PVC, CPVC, Polypropylene, or stainless steel. New venting will increase the project cost. Caution should be used when venting the new boilers as the flue temperatures may be beyond the temperature limits of the venting material. If the heating system requires 180 degree water at the outdoor design temperature, this may exclude PVC from the venting material as it is only rated for 140 degrees. I have heard of projects where the installer used PVC pipe for a hydronic boiler and the excessive flue gas temperature caused the PVC to fail and collapse. The following is a temperature rating of different venting material.

Venting Material	Temperature Limit
PVC	140 Degrees F
CPVC	180 Degrees F
Polypropylene	230 Degrees F

Sidewall venting Many new boilers can be sidewall vented. That leads to some interesting challenges. If you read the codes for sidewall venting, I dare you to meet the codes when installing a sidewall vented boiler in an older building with operable windows.

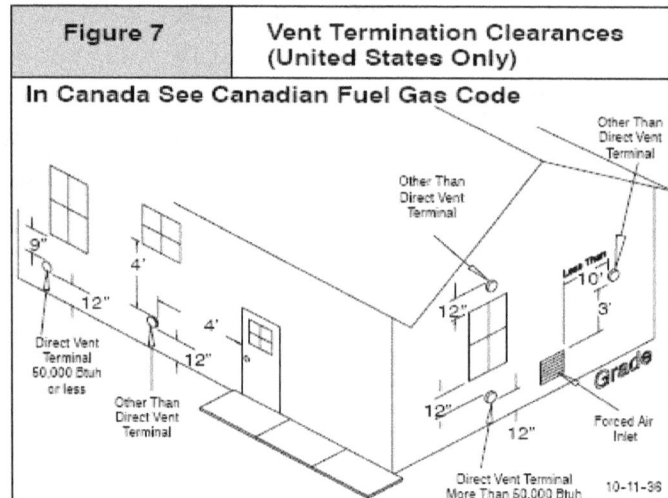

Figure 7 — Vent Termination Clearances (United States Only)
In Canada See Canadian Fuel Gas Code

- "Where adjacent to walkways, the termination of mechanical draft systems shall not be less than 7 feet above walkway"
- "3 feet above any forced air inlet within 10 feet
- 4 feet below, 4 feet horizontally from or 1 foot above any door, window or gravity vent into building
- No closer than 3 feet from an interior corner formed by 2 walls perpendicular to each other
- Not within 3 feet horizontally or directly above an oil tank or gas meter
- At least 12 inches above finished grade

A nursing home in my area had a new heating system installed that used sidewall venting. The boiler was vented through the wall into an area where two perpendicular walls met. The engineer and contractor installed the venting as per the code. The system worked well the first heating season. During the second heating season, a dangerous condition arose. During a cool rainy day, the combustion gases would not vent due to the weather conditions. Some of the flue gases gathered in the ell formed by the two perpendicular walls and was drawn into the ventilation inlets of the nursing home unit ventilators. The patients had to be evacuated. A lawsuit was filed and the engineer and contractor had to install a stainless steel vent that rose above the four-story structure at their own cost.

Boiler Startup

When designing a boiler room, part of the startup includes the adjustment of the fuel to air for the burner. In most instances, this adjustment is performed when the weather is warm. As a result, the adjustment is not correct because the boiler cycles frequently due to the loop heating too quickly in warmer weather. The proper way to do a combustion test is when the weather is cool and the boiler can operate for an extended time. The burner should operate for 15 minutes prior to adjustment of the air and fuel. This allows the flame to stabilize. I suggest that as part of the contract, the fuel to air ratio should be checked and adjusted again when the weather is cold. This will assure the client that the burner is operating at the proper efficiency.

Why is my efficiency lower than the brochure says? I purchased a new automobile a few years ago, partly because of the fuel ratings that were pasted on the window. As I drove the vehicle, I saw that my gas mileage was not what was advertised. It was about 20% lower. When I asked the dealer about this, he shrugged and said that I was not driving the vehicle in the way they did during the testing procedure. What does that have to do with a boiler? Well, it explains quite a bit as to why we do not get the same efficiency in the field that the boiler brochure touts. If we were going to start a new boiler company, prior to publishing an efficiency rating, we would have to test it to verify our claims about how efficient our boiler is. To properly certify boiler efficiency, the manufacturer must test it according to the requirements of BTS-2000 as published by the Hydronics Institute of AHRI. Let us assume that the boiler is 500,000 Btuh. The testing criteria for a boiler that size is as follows:

Flue The boiler test flue is limited to one 90 degree Ell and a single four foot piece of stack. In real life, it is rare that the flue connections consist of a flue that short.

Duration The test should be done when the boiler reaches a state of equilibrium, usually 15 minutes.

Temperatures The inlet water temperature for a standard efficiency boiler should be between 35 degrees F and 80 degrees F while maintaining an outlet temperature of 180 degrees F +- 2 degrees F. If the boiler is a condensing boiler, the inlet temperature should be 100 degrees F and the outlet should be 180 degrees F.

High fire only The test is done with the burner at high fire only. Low fire conditions are not included in the test.

If you operate the boiler at these conditions in the field, your client will be buying a new boiler sooner than later. There is movement in the industry to include low fire in the boiler testing criteria.

Combustion Analysis A modulating burner will take much longer to check and adjust than an on-off or low high low burner which increases the maintenance costs. In addition, it is almost impossible to get the same efficiency throughout the entire range. The new linkage-less controls make the setup somewhat easier. The linkage less controls are typically used on boilers larger than 150 boiler horsepower.

When adjusting the fuel to air ratio at low fire, a subtle change has taken place in the industry. Many of the burner manufacturers have found that when the burners stay on low fire for extended times, damage to the burner heads may occur. In addition, the fuel and air do not mix as well, leading to elevated carbon monoxide and emission levels. To get better mixing and better burner protection, you will now see that many burners require more air at low fire. In some instances, it may be 20-40% more at low fire. This could lower the efficiency by 1-2% more. In addition, extended operation at low fire in standard boilers could allow the flue gases to condense, destroying the boiler with category I vents.

```
          testo 327-1
V1.18            01909540/USA


12/20/2012          09:44:31

Fuel                   Nat Gas
CO2 max                 11.7 %

          Flue gas

  428.7 °F      T stack
   8.91 %       CO2
   82.9 %       EFF
   27.2 %       ExAir
    4.9 %       Oxygen
      3 ppm     CO
      4 ppm     CO AirFree
    ---  inH2O  Draft
   68.4 °F      Ambient temp
   60.1 °F      Instrum temp
    ---  °F     Diff. temp.
    ---  inH2O  Diff. Press
      0 ppm     CO Ambient
```

There are many things that affect the fuel to air ratio of the burner. It could be something as small as cleaning the blower wheel. When cleaned, it brings more air to the burner and combustion process, affecting the fuel to air ratio. When working on a boiler, it is a good idea to check the combustion at

any time that you are servicing the boiler. The linkages could have slipped or changed since the last visit. It is especially crucial if you change or adjust the following components: gas valve, blower motor, blower wheel, gas pressure.

The combustion analyzer should be calibrated yearly. This will assure that you are getting the most accurate readings. I would also suggest that you use one with a printer so that you can time and date stamp the readings and it should go into the customer file. This will protect you in case a linkage slips or someone changes the linkage after you leave. If you do not have a printer, a picture from your phone of the readings could work.

How much air? For perfect efficiency, a gas burner requires 10 parts of air for each part of fuel. For simplistic sake, let us use cubic feet. A burner will require 10 cubic feet of air for each cubic foot of gas for perfect efficiency. It is not realistic to think we can stay at that efficiency so the burners are designed to introduce "excess air" to the burner. The job of the excess air to escort the fuel burning mixture from the boiler to the outside. On industrial boilers with sophisticated controls that include an oxygen trim system, we usually see about 15% excess air. On most commercial boilers, they usually ask for 20% excess air. So our mixture would look like this:

1 cubic foot of natural gas

10 cubic feet of air for perfect efficiency

2 cubic feet of air for excess air 20%

Where to test The picture to the right shows a hole in the flue where the service company performed the combustion analysis. This was the incorrect location as it was above the draft diverter. The draft diverter introduces secondary air to the flue. The probe location would have sensed the dilution air and not the actual burner fuel to air. When testing a boiler with a draft hood, always be sure the probe is below the diverter. See arrow.

How combustion air affects boiler efficiency
When you are adjusting the fuel to air ratio of a burner, the amount of excess air can directly affect the operation of the boiler as it changes the dew point or condensing temperature of the flue gases. For example, we are taught that flue gases will condense at 140 degrees F but that only happens when the oxygen content of the flue gases is below 3%. If we have 5% oxygen content in the flue gases, our condensing temperature drops 10 degrees to 130 degrees F. The chart will show the dew point temperatures of flue gases. At the 20% excess air that many burners use, the condensing temperature of the boiler is now 131^0F. If that temperature looks familiar, it is the design temperature for European condensing hydronic systems that I referenced on page 16. Cool, huh?

Excess Air Efficiency

O2	CO2	Excess %	Dew Point
3.00%	10.0%	15.00%	133
4.00%	9.50%	20.0%	131
5.00%	9.00%	29.0%	130
6.00%	8.40%	36.0%	128
7.00%	7.90%	46.5%	123
8.00%	7.30%	56.5%	122
9.00%	6.70%	68.6%	118
10.0%	6.20%	83.5%	116
11.0%	5.60%	100%	113

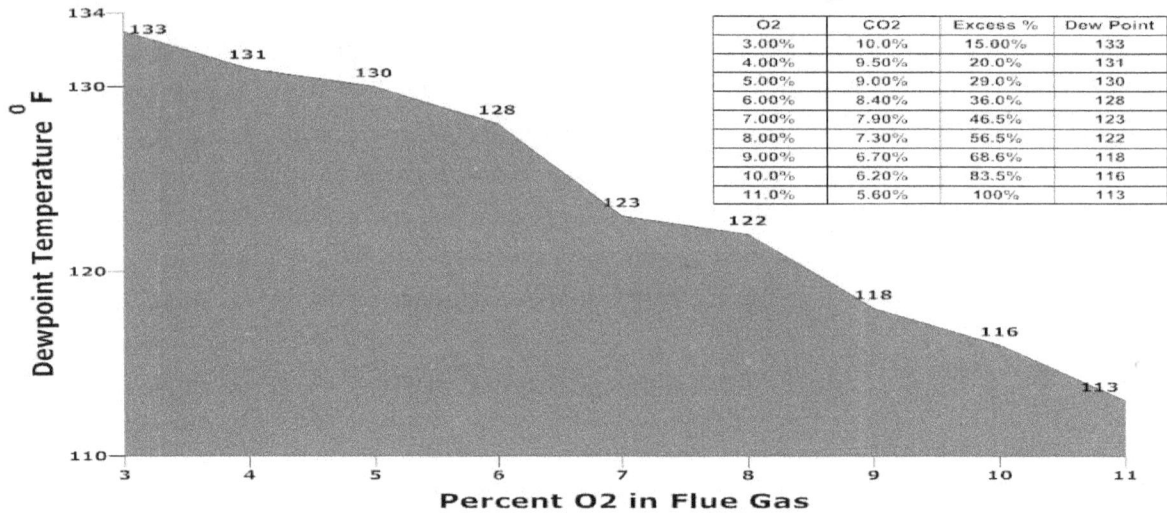

Draft should be measured and logged during the start up. Excess draft can steal the heat from the boiler and direct it up the stack. This will waste energy. It could also distort the flame pattern, causing Carbon Monoxide CO to form.

Rule of thumb. For every 0.01" w.c. the excess draft can be reduced, the fuel consumption is reduced by 1% in Category 1 appliances. On Category 4 boilers, the draft is about 1" positive.

Typical Draft Readings for Boilers		
Type of Heating System	**Overfire Draft**	**Stack Draft**
Gas, Atmospheric	Not Applicable	-.02 to -.04" WC
Gas, Power Burner	-.02" WC	-.02 to -.04" WC
Oil, Conventional	-.02" WC	-.04 to -.06" WC
Oil, Flame Retention	-.02" WC	-.04 to -.06" WC
Positive Overfire Oil & Gas	+.4 to +.6	-.02 to -.04" WC
Category 4 Positive	Positive	+1.0" w.c.

Clocking a Gas Meter

Many engineers specify that the new boilers be "clocked" to assure proper firing rate. Clocking a gas meter is a way of verifying the actual firing rate of the boiler. There are a couple of ways to perform that task. To properly "clock" a meter, be sure that your boiler is the only apparatus that is firing. The boiler should also be at high fire. A way to check this is to shut off all the items in the boiler room and observe the gas meter. If the hands on the dial are still moving, there are other items consuming gas. It is sometimes difficult to perform this task in a commercial building, as there are could be many items operating simultaneously.

Some commercial gas meters require you to compensate for the temperature as well as the pressure of the gas. When the meter was calibrated, it was at a certain temperature and gas pressure at the factory. The meter above was calibrated using 60 degree F gas. The field conditions may be different. In the end of the chapter, there are some figures that will help you to compensate for the conditions found at the job site.

How to Clock the Meter

Once you are sure that there are no other appliances using gas, you will need to start your unit and assure that it is firing at full rate.

You then want to count the number of revolutions the most sensitive dial on the gas meter makes in one minute. Most natural gas has a heating value of 1,000 BTU/cubic foot. Let us assume that our most sensitive dial is ½ cubic feet per revolution.

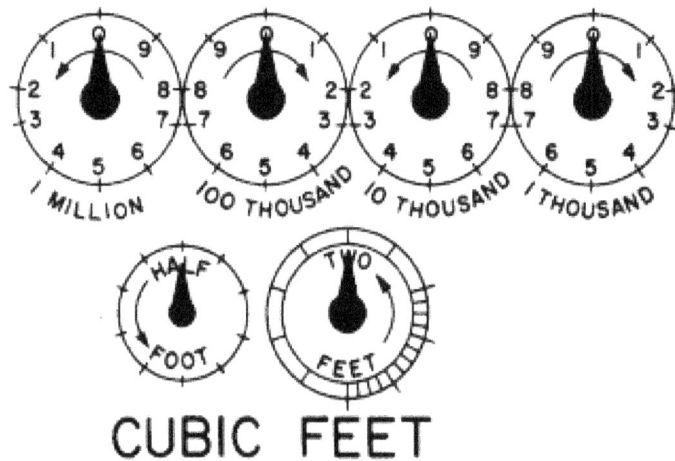

A. Count the revolutions the ½ cubic foot dial makes in one minute.
B. Multiply the revolutions by 30,000 to obtain the

firing rate in Btu's/ Hr

For example, the ½ cubic foot dial made 3.2 revolutions in one minute. The boiler is firing at 3.2 revolutions x 30,000 BTU/revolution = 96,000 BTUH. If you find that the heating value is different from 1,000 Btu's per cubic foot, you would have to make an adjustment. The local gas company could inform you of the heating content of their gas. For example, if the gas company tells you that the heating value is 1,050 BTU/ cubic foot, you would need to adjust your final reading. 1,050 BTU/ Cubic foot (Actual BTU) divided by 1,000 Btu/ Cubic foot (This was assumed to be the BTU content) = 1.05. Therefore, to recalculate the new rate, we would multiply 96,000 Btuh (From above) x 1.050 = 100,800 Btu/ HR. This is the actual firing rate of the appliance.

The 30,000 calculation only works with ½ cubic foot dial. For other size dials, see below.

Remember our basic formula is Number of revolutions x factor below = BTU/ Hr. This is based on 1,000 BTU/ Cubic Foot.

NOTE: To get a more accurate reading, it is better to allow the test to be done for a longer time. I would recommend 5 minutes. You would then divide the reading by 5 to get the average. The chart below features different timing for the dials, up to five minutes. For example, if the 5 "Cubic Feet per Revolution" dial made two revolutions in five minutes, your firing rate would be as follows:

60,000 x 2 = 120,000 Btuh

Multiplying Factor for Gas Meter

Cubic Feet per Revolution	1 Minute Timing	2 Minute Timing	3 Minute Timing	5 Minute Timing
	BTUH			
½	30,000	15,000	7,500	6,000
1	60,000	30,000	15,000	12,000
2	120,000	60,000	30,000	24,000
5	300,000	150,000	75,000	60,000
Based on 1,000 Btu per cubic foot of gas				

Clocking a Gas Meter Option 2

A second method for "clocking" a gas meter is as follows:

Start the boiler; making certain that no other gas-fired appliance is operating. Measure the amount of time it takes for the smallest dial to make one complete revolution. In the above dials, the ½ cubic foot dial is the timing dial.

Refer to a natural gas timing chart under ½ cubic foot column and see what the input is to your boiler.

Check and compare the calculated input with the input rating on the heating unit data plate. If the unit is under-fired or over-fired by more than 10%, check the gas pressure to the unit with a fluid filled manometer and adjust as necessary.

(For example, the unit being tested takes 29 seconds for the ½ cubic foot dial to make one complete revolution. Using the chart, this translates to 62 cubic feet per hour. Based upon the assumption that one cubic foot of natural gas has 1,000 BTU's (Check with your local utility for actual BTU content), the calculated input is 62,000 BTU's per hour.

You will get a better reading by allowing the dial to rotate several times and dividing the total by the amount of revolutions to get an average. In the above example, if it took 1 minute, 27 seconds or 87 seconds to make three revolutions, our average input would be 29 seconds.

Natural Gas Timing Chart in Cubic Feet/ Hour

Seconds for one revolution	1/2 Cu Ft	1 Cu Ft	2 Cu Ft	5 Cu Ft
10	180	360	720	1,800
11	164	327	655	1,636
12	150	300	600	1,500
13	138	277	555	1,385
14	129	257	514	1,285
15	120	240	480	1,200
16	112	225	450	1,125
17	106	212	424	1,059
18	100	200	400	1,000
19	95	189	379	947
20	90	180	360	900
21	86	171	345	857
22	82	164	327	818
23	78	157	313	783
24	75	150	300	750
25	72	144	288	720
26	69	138	277	692
27	67	133	267	667
28	64	129	257	643
29	62	124	248	621
30	60	120	240	600
31	58	116	232	581
32	56	113	225	563
33	55	109	218	545
34	53	106	212	529
35	51	103	205	514
36	50	100	200	500
37	49	97	195	486
38	47	95	189	474
39	46	92	185	462
40	45	90	180	450
40	45	90	180	450

Seconds for one revolution	1/2 Cu Ft	One Cu Ft	Two Cu Ft	Five Cu Ft
41	44	88	176	440
42	43	86	172	430
43	42	84	167	420
44	41	82	164	410
45	40	80	160	400
46	39	78	157	391
47	38	77	153	383
48	37	75	150	375
49	37	73	147	367
50	36	72	144	360
51	35	71	141	353
52	35	69	138	346
53	34	68	136	340
54	33	67	133	333
55	33	65	131	327
56	32	64	129	321
57	32	63	126	316
58	31	62	124	310
59	30	61	122	305
60	30	60	120	300
62	29	58	116	290
64	29	56	112	281
66	29	54	109	273
68	28	53	106	265
70	26	51	103	257
72	25	50	100	250
74	24	48	97	243
76	24	47	95	237
78	23	46	92	231
80	22	45	90	225
82	22	44	88	220
84	21	43	86	214
86	21	42	84	209
88	20	41	82	205

NOTES:

On a commercial gas meter, you may have to calculate a pressure and/or temperature correction factor. You will need to contact the local gas company for this factor.

Gas Pressure Correction Factor

	Actual Meter Pressure (psi)		Meter Base Pressure		
	4 oz or 7" w.c.	8oz or 14" w.c.	10 oz or 17.5" w.c.	1 Psi or 28" w.c.	2 psi or 56" w.c.
0	0.983	0.966	0.958	0.935	0.878
¼	1.0	0.983	0.975	0.951	0.893
½	1.017	1	0.992	0.968	0.909
5/8	1.026	1.008	1	0.976	0.916
1	1.051	1.034	1.025	1.000	0.929
2	1.119	1.101	1.092	1.065	1.000
3	1.188	1.168	1.158	1.130	1.061
4	1.256	1.235	1.225	1.195	1.122
5	1.324	1.302	1.291	1.260	1.183
6	1.392	1.369	1.358	1.325	1.244
7	1.461	1.436	1.424	1.390	1.305
8	1.529	1.503	1.491	1.455	1.366
9	1.597	1.570	1.557	1.520	1.427
10	1.666	1.638	1.624	1.584	1.488

Gas Temperature Correction Factor

Gas Temperature Degrees F	Meter Calibration Temperature				
	60 Degrees F	65 Degrees F	68 Degrees F	70 Degrees F	72 Degrees F
0	1.130	1.141	1.148	1.152	1.157
5	1.118	1.129	1.135	1.140	1.144
10	1.106	1.117	1.123	1.128	1.132
15	1.095	1.105	1.112	1.116	1.120
20	1.083	1.094	1.100	1.104	1.108
25	1.072	1.082	1.089	1.093	1.097
30	1.061	1.071	1.078	1.082	1.086
35	1.051	1.061	1.067	1.071	1.075
40	1.040	1.050	1.056	1.060	1.064
45	1.030	1.040	1.046	1.050	1.053
50	1.020	1.029	1.035	1.039	1.043
55	1.010	1.019	1.025	1.029	1.033
60	1.000	1.010	1.015	1.019	1.023
65	0.990	1.000	1.006	1.010	1.013
70	0.981	0.991	0.996	1.000	1.004
75	0.972	0.981	0.987	0.991	0.994
80	0.963	0.972	0.978	0.981	0.985
85	0.954	0.963	0.969	0.972	0.976
90	0.945	0.955	0.960	0.964	0.967
95	0.937	0.946	0.951	0.955	0.959
100	0.929	0.938	0.943	0.946	0.950

Gas Pipe Line Sizing

Steel Pipe Size	Pipe Length			
	10 Feet	20 Feet	40 feet	80 Feet
	Capacity in Cubic Feet per hour			
½"	120	85	60	42
¾"	272	192	136	96
1"	547	387	273	193
11/4"	1,200	849	600	424
1 ½"	1,860	1,316	930	658
2"	3,759	2,658	1,880	1,330
2 ½"	6,169	4,362	3,084	2,189
4"	23,479	16,602	11,740	8,301

Each cubic foot of gas roughly equals 1,000 Btuh

Some Notes on Gas Piping

- ✓ Gas fittings should be malleable iron and not black iron.
- ✓ Many gas companies do not permit bushings to be installed in gas lines. They prefer bell reducers.
- ✓ Threaded fittings greater than 4" shall not be used except where approved. IFGC 403.10.4
- ✓ Piping shall not be installed in or through a ducted supply, return or exhaust, or a clothes chute, chimney or gas vent, dumbwaiter or elevator shaft. IFGC 404.1
- ✓ Piping in concealed locations shall not have unions, tubing fittings, right and left couplings, bushings, compression couplings and swing joints made by a combination of fittings IFGC 404.3

IFGC = International Fuel Gas Code

Gas Pipe Threads

The following table shows the length and approximate number of threads on gas lines.

Iron Pipe Size (Inches)	Approximate Length of Threaded Portion	Approximate number of threads to be cut
1/2	¾	10
¾	¾	10
1	7/8	10
1 ¼	1	11
1 ½	1	11
2	1	11
2 1/2	1 ½	12
3	1 ½	12
4	1 5/8	13

Glycol for the System

If you are contemplating the use of glycol in your hydronic system, there are several important factors to consider.

Can Your Boiler Tolerate Glycol? Some boilers, such as aluminum ones, do not react well with standard glycol. Before installing glycol in your system, check with the boiler manufacturer. Glycol designed for aluminum boilers can be ordered. Some boiler manufacturers have specific requirements when adding glycol for their boilers.

Reduced Efficiency When adding glycol to a hydronic heating system, it will reduce the heat transfer efficiency of the system. For example, when you add 20% propylene glycol to the system, the heating system efficiency drops 3% lower. At 50% concentration, your capacity drops 10%. The 80% boiler suddenly dropped to a little over 70% efficient. This should be factored in when sizing the new system. In addition, the compression tank will have to be oversized by 20% to compensate for the glycol. The pumps also lose about 10% efficiency when the system is filled with a 50% mixture of glycol and water.

Increased Maintenance Systems containing glycol require extra maintenance. The pH has to be checked yearly. The glycol composition has to be checked using a refractometer. Glycol will breakdown eventually and may cause damage to your system. If the concentration level drops in your tests, this should indicate a piping leak. The leak should be found and repaired.

Cleaning the system The system should be cleaned prior to introduction of glycol. Flush the system with a heated 1-2% solution of Trisodium phosphate for 2 to 4 hours, then drain and rinse thoroughly. This will remove excess pipe dope, cutting oils and solder flux.

Automotive Glycol? This is not recommended as it was not designed to be used in a hydronic system. Automotive antifreeze is formulated with silicates, which tend to gel, reducing heat transfer efficiency. Use an inhibited glycol designed for heat transfer applications.

Which Type of Glycol? You can choose between ethylene and inhibited propylene glycol. Uninhibited glycol is very corrosive and could lead to damage in your system. Ethylene glycol is toxic to humans and animals. A special permit may be required when using it. Ethylene glycol will provide slightly better freeze protection than propylene glycol. Propylene glycol is more environmentally friendly and not toxic.

How Much Glycol? The most common percentage is between 20% to 50% concentration of glycol.

Water Quality Water chemistry is a concern when you introduce glycol into your system. Poor water quality can lead to scale, sediment deposits, or the creation of sludge in the heat exchanger which will reduce heat transfer efficiency. It can damage the system by depleting the corrosion inhibitor and promoting a number of corrosions including general and acidic attack corrosion. Before using the local water, it should be analyzed by an water treatment expert. Good quality water contains:

- Less than 50 ppm of calcium
- Less than 50 ppm of magnesium
- Less than 100 ppm (5 grains) of total hardness
- Less than 25 ppm of chloride
- Less than 25 ppm of sulfate

The glycol should be mixed at room temperature with either demineralized or deionized water if the water quality is questionable.

The safest, although not the least expensive, solution may be to order premixed glycol from the manufacturer.

Freezing Point

Concentration by volume	Ethylene Glycol	Propylene Glycol
50%	-37F	-28F
40%	-14F	-13F
30%	+2F	+4F
20%	+15F	+17F

Maintenance The glycol must be checked at least once a year in accordance with the manufacturer's recommendations. A base line analysis should be performed within two to four weeks of initial mixing. This measurement will be used to verify that the fill was completed properly, and will serve as a reference point for comparison with future test results. As a bare minimum, the solution should be analyzed for glycol concentration, solution pH and general fluid quality.

Solution Testing If you are using a 30% to 50% solution, the pH should be between 8.3 and 9.0. If the level falls below 8.0, this could indicate lowered inhibitors. In some instances, inhibitors can be added. If the pH falls below 7.0, the glycol should be removed and flushed. This level of pH could cause damage to the boiler and piping. The system should be tested once a month.

Mark the System When you are done with the system and turn it over to the owner, the system should have clear signage that tells anyone working on the system that it contains glycol. It should also have the following information:

- Type of glycol that is used in the system, ethylene or propylene
- Concentration of glycol
- Initial pH readings
- Installation date of glycol
- Last test date

Some Glycol Considerations Galvanized pipe should not be used in the hydronic system as the Zinc could have an adverse effect and form sludge.

Make sure the system is clean before filling. Pre-fill flushing is highly recommended.

Mix the solution at room temperature.

In order to minimize the possibility of glycol loss due to undetected leaks, hydrostatically test the system for 24 hours prior to filling.

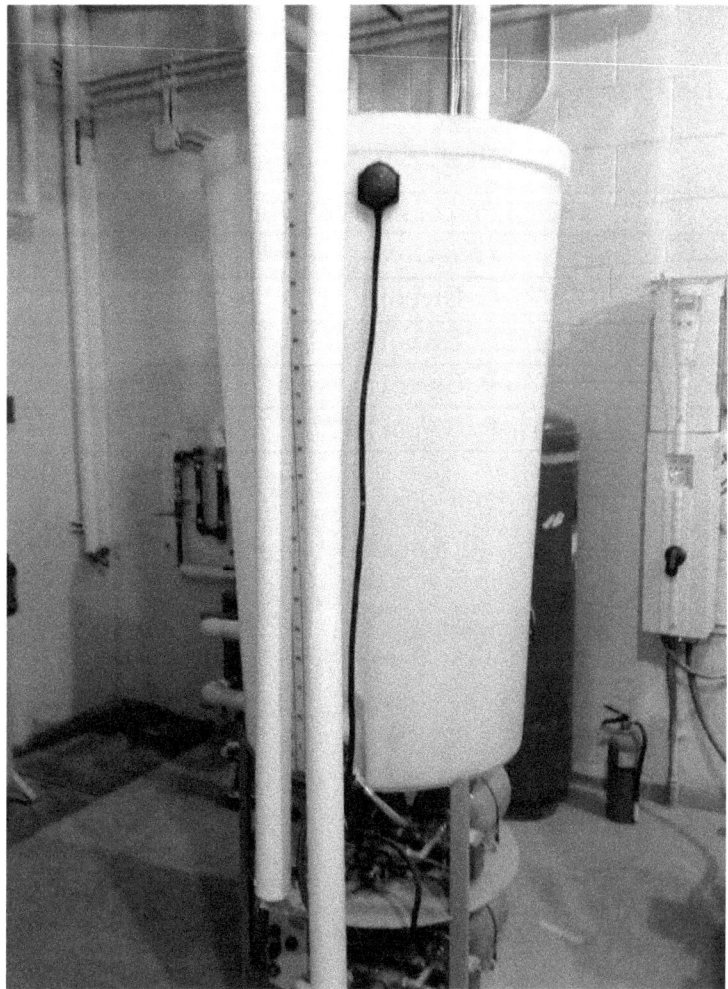
Glycol Tank

Never use a chromate water treatment in a system with glycol. The chromate will damage the glycol and can lead to severe system degradation.

Do not use in a system that may have a solution temperature over 300F.

Do not use check valves or closed zone valves that would isolate a part of the system, preventing proper expansion and resulting in freeze damage.

A strainer, sediment trap, or some other means for cleaning the piping system must be provided. It should be located in the return line ahead of the boiler and pump. This must be cleaned frequently during initial operation.

Automatic make-up water systems should be avoided in order to prevent undetected dilution or loss of glycol.

121

Check local codes to see if systems containing these solutions must include a back-flow preventer, or an actual disconnect from city water lines.

Do not use glycol in steam systems.

BTU per Hour with Glycol

Glycol Percentage & Type	Formula
No Glycol	BTUH = GPM x 500 x $\Delta t\ ^0F$
30% E. Glycol @ 68 ^0F	BTUH = GPM x 445 x $\Delta t\ ^0F$
50% E. Glycol @ 32 ^0F	BTUH = GPM x 395 x $\Delta t\ ^0F$
30% P. Glycol @ 68 ^0F	BTUH = GPM x 465 x $\Delta t\ ^0F$
50% P. Glycol @ 32 ^0F	BTUH = GPM x 420 x $\Delta t\ ^0F$

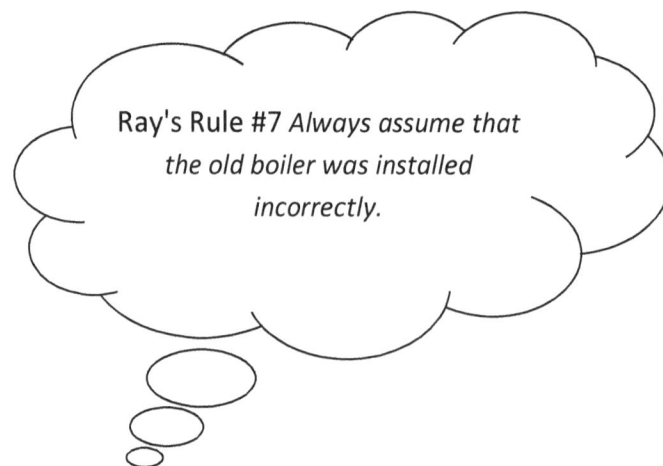

Ray's Rule #7 *Always assume that the old boiler was installed incorrectly.*

Don't forget these

A water meter is a good idea to use on any boiler project. It allows the owner to monitor leaks in the system. A water sensor will alarm if water is detected on the floor.

Water Meter	Water Sensor

Water treatment The new boiler will require water treatment and a treatment specialist should be consulted. The construction of the new boiler should be considered as well as the system piping. If the new boiler is cast aluminum, the water treatment will be different than if it is stainless steel or copper.

Chemical Pot Feeder	Chemical Injection Pump

Some cast iron boilers with the rubber seals between the sections have to be careful of the water treatment as it could affect the seals. On hydronic systems, you typically will see a pot feeder that is installed in a sidearm fashion. Most chemical treatments prefer this type of feeder because they use an inclusive type of chemical treatment, an all in one treatment. A new type of system is being used instead of the traditional pot feeder. It is called a Feeder Filter. It looks like the pot feeder but it has an internal filter bag that is used to filter dirt from the system. The internal filter bags can filter particles down to 1 micron in size.

Steam systems do not typically use this type of chemical feed system. You are probably going to use chemical injection pumps that will be fed into the boiler feed tank. Some chemicals such as amines, are fed directly into the steam header. In any event, the chemical levels should be checked on a regular basis.

Carbon Monoxide Detector I urge all my clients to install a carbon monoxide sensor and alarm in the boiler room. Carbon Monoxide or CO is odorless and colorless and difficult to detect without a test instrument. Combustible gas detectors are also a great idea for inside a boiler room. If using a combustible gas detector, remember that natural gas is lighter than air and will rise. Propane is heavier than air and will fall.

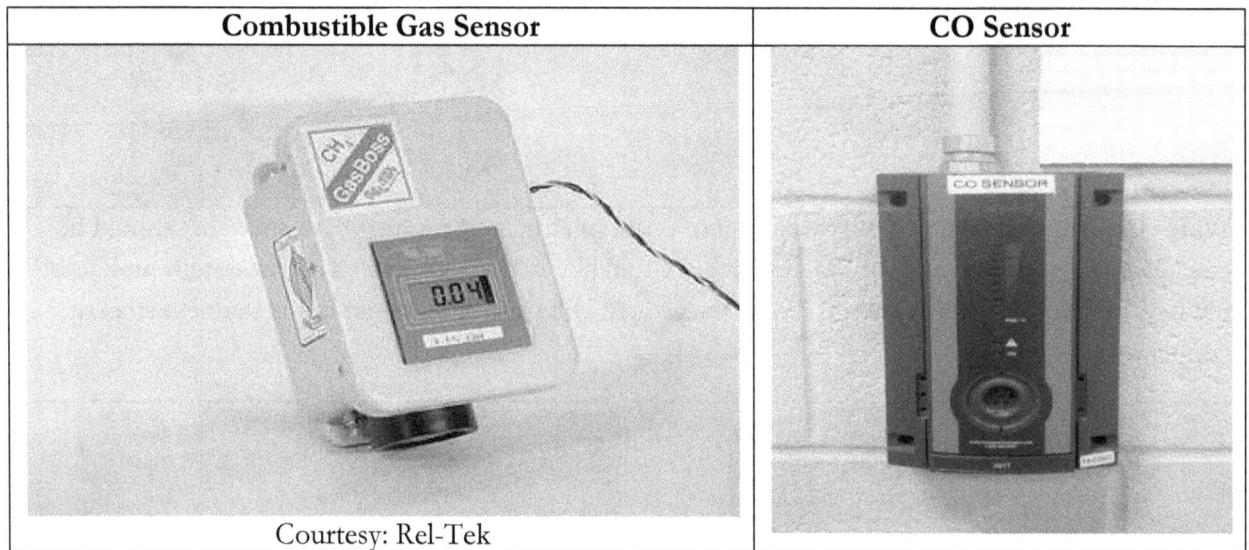

Combustible Gas Sensor	CO Sensor
Courtesy: Rel-Tek	

Thermometers are a great idea to install on a hydronic project. They help the service technician to trouble shoot the system. I would install them on the supply and the return pipes. Another great idea is to install a thermometer in the stack from the boilers. Not only will it help when servicing the boiler but the hole can be used for the combustion analyzer probe. Be care when installing one on condensing boilers as the vent may be pressurized.

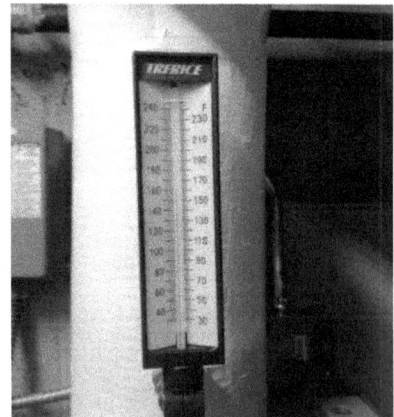

Cleaning the Boiler

Steel or Cast Iron Hydronic Boilers Consult with the boiler manufacturer about the products they recommend. In most instances, Tri Sodium Phosphate or TSP is mixed with the water and circulated for 2-3 hours. Drain, flush and refill with fresh water. After flushing the cleaner from the system, a pH test should be done to verify that the water is within the proper range. The typical pH range is 7.0 - 8.5. Please check with the manufacturer to verify their requirements. Some cast iron boilers have special requirements due to the material that is used between the sections.

One boiler manual that I saw has the following instructions for cleaning a new steam boiler.

Use equal parts of

Trisodium Phosphate, Caustic Soda, Soda Ash

It recommends using 1/2 pound per boiler horsepower. It suggests operating the steam boiler, without allowing it to steam for 16-24 hours. It also suggests allowing the condensate to be wasted to the drain for 7-10 days and checking the water treatment chemical levels every four hours. See, you really need to read the owner's manual prior to providing a firm price. This startup lasted for days.

> **Note: If you have an aluminum water boiler, you may not be able to use Tri Sodium Phosphate (TSP) as a cleaning agent for the boiler. TSP has a high pH level and could remove the natural protective oxide layer from the aluminum.**

Cleaning The Pipes A client of mine purchased pipe cleaning chemicals from his cleaning company. The sales person assured my client that the chemical would remove all the old rust and scale from the boiler and piping, resulting in lower energy costs. Once he installed the chemicals in the piping, leaks sprung up everywhere. The customer had to replace almost a mile of pipe and two old boilers. It was an expensive lesson. Always consult a water treatment professional.

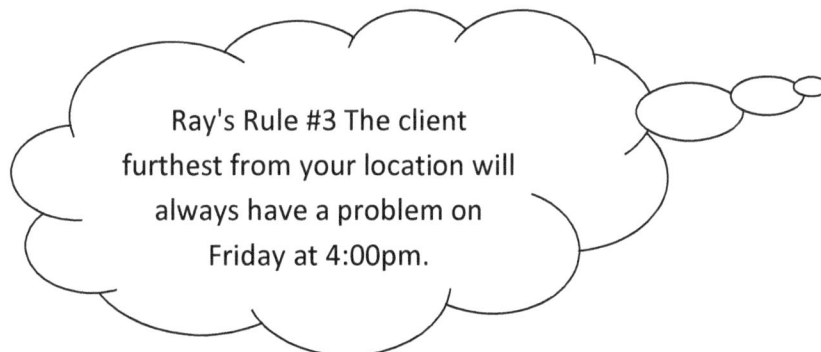

Ray's Rule #3 The client furthest from your location will always have a problem on Friday at 4:00pm.

What about Service?

When you design and or install a commercial boiler room, the components will require regular service and maintenance. If your state follows ASME CSD1, it is required:

ASME CSD1 - American Society of Mechanical Engineers Controls and Safety Devices for Automatically Fired Boilers requires that boilers to be tested and checked yearly. The following are included with CSD1 to assure the boilers are maintained properly.

"The owner of an automatic boiler system shall develop and maintain a formal system of periodic preventive maintenance and testing."

"Tests shall be conducted …results recorded in boiler log or in the maintenance record or service invoice."

"Any defects found shall be brought to the attention of the boiler owner and shall be corrected immediately."

Who will be doing the service? If you are designing or installing a new systems, chances are that the system will be much more complex than the old system. I would urge that you have a thorough instruction period for the onsite personnel and it should be recorded so they will remember it next fall or if someone new gets hired. Many facilities do not have money for maintenance. I would recommend a multi-year service agreement included in the proposal.

Courtesy: Monastery of the Blessed Sacrament

Proprietary Parts "If your ear is itchy, that means that someone is talking about you" my superstitious Irish mother used to say. If you have ever designed or installed a heating system with proprietary parts, I would wager that your ears have been itchy. The service department of my company hears the complaints from the upset building owners that the parts they need are not stocked or are much more expensive than standard off the shelf controls. If the systems has proprietary parts, add some of the more common components that are apt to fail like a flame sensor or igniter into the boiler bid.

Trouble Shooting Commercial Boilers

Most No Heat calls on a power burner happen during pilot sequence.

Typical sequence of operation on a power burner

There is a call for heat by a control.

The blower starts and the operation is verified by the sensing control. This is usually a pressure switch in or on the burner control panel.

Once blower movement is detected, the burner will enter the pre purge phase. This will be as long as it takes for 4 air changes in the boiler. This is usually between 30 seconds to 3 minutes. If it is a modulating burner, it will drive to the high fire position and satisfy the High Fire Start Switch before pre purge timer starts.

After pre purge phase, the modulating motor will drive to low fire position and satisfy the Low Fire Start Switch, usually in the modulating motor.

Once low fire position is verified, the transformer will send a spark to the ignition electrode. Be careful as the voltage required for the spark is between 7,000 and 10,000 volts. At the same time, the pilot solenoid valve will open, allowing fuel into the burner. The flame safeguard will verify that the pilot is ignited. If the flame safeguard does not sense a flame, it will go into flame failure. This requires pushing the reset button on the front of the flame safeguard. The flame signal strength may be checked using a multi meter.

If the pilot flame is verified, the flame safeguard will send a signal to the main gas valve, opening it. The flame safeguard will monitor the flame strength. If it falls below the setting, the flame safeguard will shut off the burner and will result in a flame failure. The flame will stay on until the call for heat is ended.

Some old burners had a post purge which operated the blower after the call for heat ended to purge any leftover combustibles from the boiler. It is not used as much as it takes heat from the boiler and sends it up the stack.

Typical Boiler Service Call

Items that should be checked during a boiler service call

Fuel Shutoff, Manual	Refractory
Fuel Shutoff, Electric	Burner Head
Pilot Assembly	Blower Wheel
Igniter	Blower Motor
Gas Pressure Regulator	Relief Valve
Gas Train Venting	Relief Valve Discharge Piping
Wiring Connections	Transformer
Control Board	Wiring
Flame Signal	Gauges & Thermometers
Modulating Motor	Draft Controls
Modulating Control	Induced Draft Fan
Low Fire Start Switch	Flue & Chimney
High Fire Start Switch	Bearings & Linkages
Flame Safeguard	Firing Rate Valve
Flow Switch	**System Items to Check**
Low Gas Pressure Switch	Backflow Preventer
High Gas Pressure Switch	Combustion Air
Operating Control	Compression tank
Limit Control	System Pressure
Low Water Cutoff	Water Feeder
Air Fuel Safety Switch	Circulating Pumps
Main Burner(s)	Emergency Door Switch
Combustion adjustment should be done during cold weather	

Can We Use the Boiler to Heat the Pool?

The first thing you should do is to calculate the number of gallons in the pool.

Estimate pool volume:

Multiply Length x Width x Average depth in Feet x 7.48 (Gallons per cubic feet) For example, if our pool is 40 feet long x 20 feet wide x 5 feet deep , the volume of water inside the pool would be 29, 920 gallons. Now, calculate the cold startup requirement. For instance, you want to raise the water in the pool from 55 degrees F to 75 degrees F.

Multiply volume x degree rise x 8.34 (weight of a gallon of water)

29,920 x 20 x 8.34 = 4,990,656 The next step is to divide the Btu's required to heat the pool to the desired temperature over the period of time we are willing to wait. If we are willing to wait 24 hours, our boiler would have to have a rated output of 207,944 Btuh. If we can wait for 48 hours, the boiler minumum output rating would be 103,972 Btuh.

During the time that the boiler is heating, our pool surface area will give up some of its heat to the ambient air. That has to be included in our calculations as well. The following are charts that show the Btu losses at various wind speeds and temperature differences.

Heat Loss from pool surface
Based on 3.5 MPH wind velocity

Temperature Difference Deg F	Btuh loss / Sq Ft Surface
10	105
15	158
20	210
25	263
30	368

Heat Loss from pool surface
Based on 5 MPH wind velocity

Temperature Difference Deg F	Btuh loss / Sq Ft Surface
10	131
15	198
20	263
25	329
30	460

Heat Loss from pool surface
Based on 10 MPH wind velocity

Temperature Difference Deg F	Btuh loss / Sq Ft Surface
10	210
15	316
20	420
25	526
30	736

Boiler Emissions

Due to worries about climate change, boiler emissions are a growing concern. The following are some of the by-products of fuel burning.

Sulfur Dioxide Sulfur Dioxide is also a by-product of combustion. It is rare with gaseous fuels. It is also a primary contributor to acid rain, which causes acidification of streams and lakes. It is released primarily from burning fuels that contain sulfur (such as coal, oil, and diesel fuel). Sulfur dioxide, when combined with water, forms sulfuric acid.

Particulate Matter Particulate matter, or otherwise known as soot, is mostly formed from incomplete combustion of fuel. It is comprised of unburned fuel, organic chemicals, soil, dust, sulfates, nitrates, oxides and or carbons. There are two basic types of particulate matter, Pm and PM_{10}. PM_{10} is particulate matter that is less that 10 microns in diameter. To see how small this is, consider that a human hair is about 70 microns. The PM_{10} is small enough to bypass the human body's natural filtering system and imbed themselves in the lungs. It can trigger asthma attacks, coughing and acute bronchitis. It is more prevalent in oil than gas. There are some areas that are trying to test for particulates as small as 2.5 microns.

Carbon Dioxide Carbon dioxide, when combined with water, forms carbonic acid.

Nitrogen Compounds or NO_x NO_x is principally made up of two components, Nitric Oxide(NO) and Nitrogen Dioxide(NO_2). NO_2, when combined with other pollutants, such as Volatile Organic Compounds (VOCs), is believed to form O3 or ground level ozone and acid rain. NO_x emissions become more prevalent when the flame temperature is above $2,800^0F$ and the fuel to air combustion ratios are between 5-7% O_2. This is typically where the commercial burner manufacturers direct the fuel to air ratios to be set.

There are two types of NOx, Thermal NOx and Fuel NOx. Thermal NOx is the most common. It is produced in the boiler when oxygen and nitrogen combine under elevated temperatures. Fuel NOx rarely occurs when firing with gas. It is common in heavier fuel oils.

The flue gases from a fossil-fueled boiler contain the following: oxygen, carbon dioxide, carbon monoxide, sulfur dioxide and free water. If allowed to condense, these acids will destroy the stack, chimney and boiler.

Pollution Conversions

After you finish the combustion analysis on your new boiler, you may be asked to forward an emissions report. To calculate that, take the readings and follow the following formulas. To convert PPM to any of the units below, multiply PPM by the number in the correct column and row.

Multiply PPM by factor below					
Fuel	Pollutant	Lb/MBTU	MG/NM3	MG/KG	G/GJ
Nat Gas	CO	0.00078	1.249	12.647	0.338
Nat Gas	NOx	0.00129	2.053	20.788	0.556
Nat Gas	SO2	0.00179	2.857	28.949	0.775
Oil #2, #6	CO	0.00081	1.249	15.118	0.354
Oil #2, #6	NOx	0.00134	2.053	24.850	0.582
Oil #2, #6	SO2	0.00186	2.857	34.605	0.811

Definitions

Lb/MBTU = pounds of pollutants per million BTU

Mg/NM3 = Milligrams of pollutants per Million BTU

MG/KG = Milligrams of pollutants per Kilogram of fuel burned

G/GJ = Grams of pollutant per Gigajoule

Based on 3% excess Oxygen and dry gas

Good luck when marrying the new equipment with the existing system.

Estimated Combustion Efficiency Tables

Natural Gas

Excess Air	O2%	CO2%	Net Stack Temperature						
			200	250	300	350	400	450	500
8.5	2.0	10.7	85.4	84.2	83.1	81.9	80.8	79.6	78.4
12.1	2.5	10.4	85.3	84.1	83.0	81.8	80.3	79.4	78.2
15.0	3.0	10.1	85.2	84.0	82.8	81.6	80.1	79.2	77.9
18.0	3.5	9.8	85.1	83.9	82.6	81.4	79.9	78.9	77.6
21.1	4.0	9.6	85.0	83.7	82.5	81.2	79.7	78.7	77.4
24.5	4.5	9.3	84.8	83.6	82.3	81.0	79.4	78.4	77.1
28.1	5.0	9.0	84.7	83.4	82.1	80.8	79.2	78.1	76.7
31.9	5.5	8.7	84.6	83.3	81.9	80.6	78.9	77.8	76.4
35.9	6.0	8.4	84.4	83.1	81.7	80.3	78.6	77.5	76.0
40.3	6.5	8.2	84.3	82.9	81.5	80.0	78.3	77.1	75.6
44.9	7.0	7.9	84.1	82.7	81.2	79.7	77.9	76.7	75.2
49.9	7.5	7.6	84.0	82.5	80.9	79.4	77.6	76.3	74.8
55.3	8.0	7.3	83.8	82.2	80.7	79.1	77.2	75.9	74.3

Estimated Combustion Efficiency Tables

#2 Fuel Oil

Excess Air	O2%	CO2%	Net Stack Temperature						
			200	250	300	350	400	450	500
9.9	2.0	14.1	89.6	88.5	87.4	86.3	85.2	84.1	82.9
12.6	2.5	13.8	89.5	88.4	87.3	86.2	85.0	83.9	82.7
15.6	3.0	13.4	89.4	88.3	87.1	86.0	84.8	83.6	82.4
18.7	3.5	13.0	89.3	88.2	87.0	85.8	84.6	83.4	82.2
22.0	4.0	12.6	89.2	88.0	86.8	85.6	84.4	83.1	81.9
25.5	4.5	12.3	89.1	87.9	86.6	85.4	84.1	82.8	81.6
29.2	5.0	11.9	89.0	87.7	86.4	85.1	83.9	82.6	81.2
33.2	5.5	11.5	88.8	87.5	86.2	84.9	83.6	82.2	80.9
37.4	6.0	11.2	88.7	87.3	86.0	84.6	83.3	81.9	80.5
41.9	6.5	10.8	88.5	87.1	85.8	84.4	83.0	81.5	80.1
46.8	7.0	10.4	88.3	86.9	85.5	84.1	82.6	81.2	79.7
52.0	7.5	10.0	88.2	86.7	85.2	83.8	82.3	80.7	79.2
57.6	8.0	9.7	88.0	86.5	84.9	83.4	81.9	80.3	78.7

Boiler Types

Old Cast Iron Sections

Copper Boilers

Water Tube Boiler Cutaway
Courtesy of Rite Boiler

Water tube

Hybrid System
Standard efficiency boiler on left

Wall Mounted Boiler

Industry Links that I like

American Boiler Manufacturers Association	www.abma.com
American Gas Association	www.aga.org
American Society of Mechanical Engineers	www.asme.org
Appropriate Designs John Siegenthaler	www.hydronicpros.com
ASHRAE	www.ashrae.org
Engineered Systems Magazine	www.esmagazine.com
Fire & Ice *(My company)*	www.fireiceheat.com
Foley Mechanical	www.foleymechanical.com
FW Behler Dave Yates	www.fwbehler.com
Healthy Heating Robert Bean	www.healthyheating.com
Heating Help Dan Holohan	www.heatinghelp.com
Johnson Controls	www.johnsoncontrols.com
Mechanical Hub	www.mechanical-hub.com
National Board of Boiler and Pressure Vessel Inspectors	www.nationalboard.org
Oil & Energy Service Professionals OESP	www.thinkoesp.org
Plumbing & Mechanical Magazine	www.pmmag.com
PM Engineer Magazine	www.pmengineer.com
Radiant Professionals Alliance	www.radiantprofessionalsalliance.org
Rel-Tek	www.rel-tek.com
Rite Boiler	www.riteboiler.com
RSES	www.rses.org
Siemens	http://w3.usa.siemens.com/buildingtechnologies/us
Triad Boiler Systems	www.triadboiler.com

Heating Formulas & Rules of Thumb

These are rules of thumb. Please check with the equipment manufacturer for actual requirements.

Combustion Air Openings

Each fuel-burning piece of equipment requires combustion air to operate safely. The following are some guidelines to help you see whether the existing combustion air louvers will be adequate for the replacement project if you are directly venting the flues.

Number of openings required = 2
Each boiler room should have two openings. One should be within one foot of ceiling and the other opening within one foot of the floor.

Size of Direct Openings
1" Free area for each 4,000 Btuh

Horizontal Openings
1" Free space for each 2,000 Btuh

Vertical Openings
1" Free space for each 4,000 Btuh

Mechanical Ventilation
1 cfm per 2,400 Btuh

Direct vent - Power Flame recommends sizing the duct for a pressure drop of 0.1" w.c. including all screens, filters, and fittings.

Combustion Air Fan Sizing			
BTUH	CFM	**BTUH**	CFM
400,000	167	1,300,000	542
500,000	208	1,400,000	583
600,000	250	1,500,000	625
700,000	292	1,600,000	667
800,000	333	1,700,000	708
900,000	375	1,800,000	750
1,000,000	417	1,900,000	792
1,100,000	458	2,000,000	833
1,200,000	500	2,100,000	875

Direct Vent Combustion Air Sizing

If you are directly venting combustion air for the boiler, this may help you when sizing the air duct. The chart below is based on using 15 cubic feet of air for every cubic foot of gas burned or about 50% excess air. Most new boilers will use about 12 parts air or 20% excess air. You will have to verify that with the boiler manfacturer.

Estimated Combustion Air Required @ Various Boiler Inputs			
Boiler Btuh Input	Cu Ft Gas / Hr**	Cu Ft Gas/ Minute	**CFM Air***
200,000	200	3.33	50
300,000	300	5.0	75
400,000	400	6.67	100
500,000	500	8.33	125
600,000	600	10.0	150
700,000	700	11.67	175
800,000	800	13.33	200
900,000	900	15.0	225
1,000,000	1,000	16.67	250
1,500,000	1,500	25.0	375
2,000,000	2,000	33.33	500
3,000,000	3,000	50.0	750
4,000,000	4,000	66.67	1,000
5,000,000	5,000	83.33	1,250
6,000,000	6,000	100.0	1,500
7,000,000	7,000	116.7	1,750
8,000,000	8,000	133.3	2,000
9,000,000	9,000	150.0	2,250
10,000,000	10,000	166.7	2,500
* Based upon 15 cubic feet of air for every cubic foot of gas burned.			
** Based on 1,000 Btu per cubic foot of gas			

How Excess Combustion Air Affects Boiler Condensing Temperatures			
O2%	CO2%	Excess Air %	Dew Point ^{0}F
3.0%	10.0%	15.0%	133^{0}F
4.0%	9.50%	20%	131^{0}F
5.0%	9.0%	29.0%	130^{0}F
6.0%	8.40%	36.0%	128^{0}F
7.0%	7.9%	46.5%	123^{0}F
8.0%	7.3%	56.5%	122^{0}F
9.0%	6.7%	68.6%	118^{0}F
10.0%	6.2%	83.5%	116^{0}F
11.0%	5.6%	100%	113^{0}F

Excess Air Efficiency

O2	CO2	Excess %	Dew Point
3.00%	10.0%	15.00%	133
4.00%	9.50%	20.0%	131
5.00%	9.00%	29.0%	130
6.00%	8.40%	36.0%	128
7.00%	7.90%	46.5%	123
8.00%	7.30%	56.5%	122
9.00%	6.70%	68.6%	118
10.0%	6.20%	83.5%	116
11.0%	5.60%	100%	113

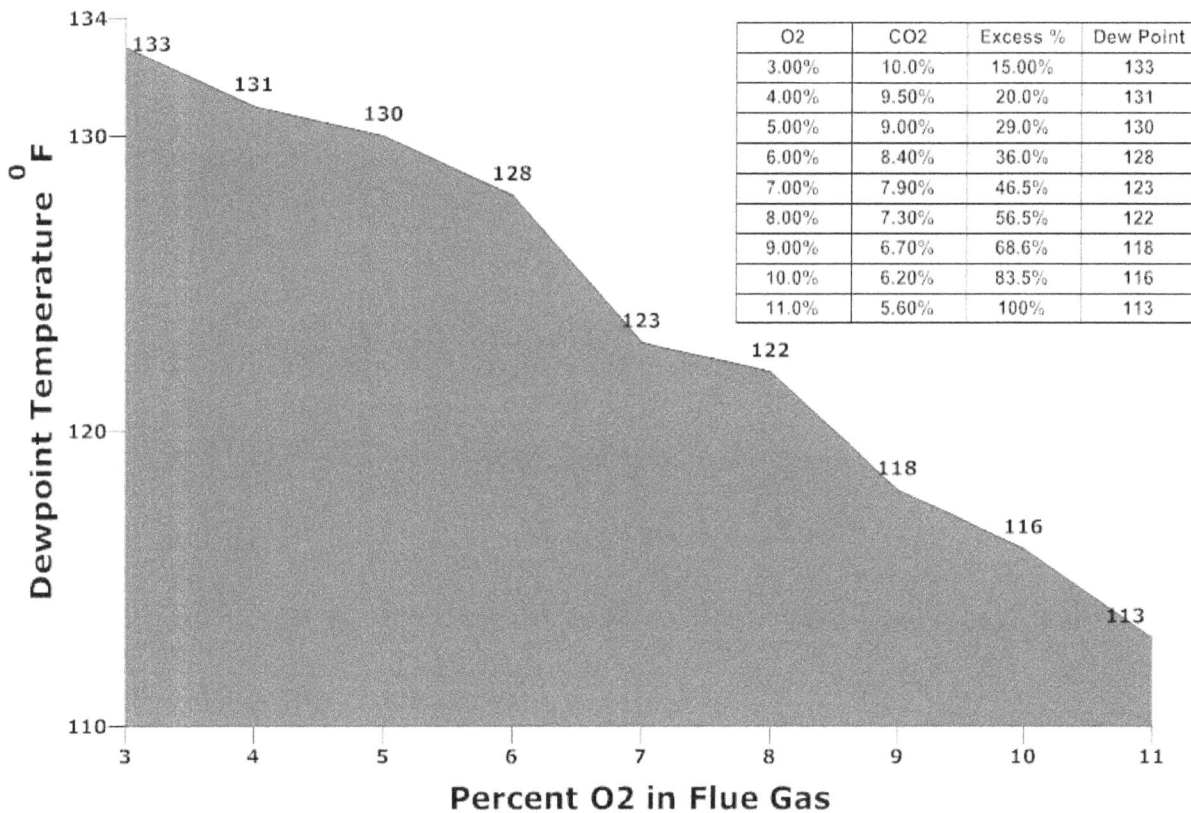

Dewpoint Temperature ^{0}F

Percent O2 in Flue Gas

Sizing Gas Train Manifold Vent
Source: Philadelphia Gas Works

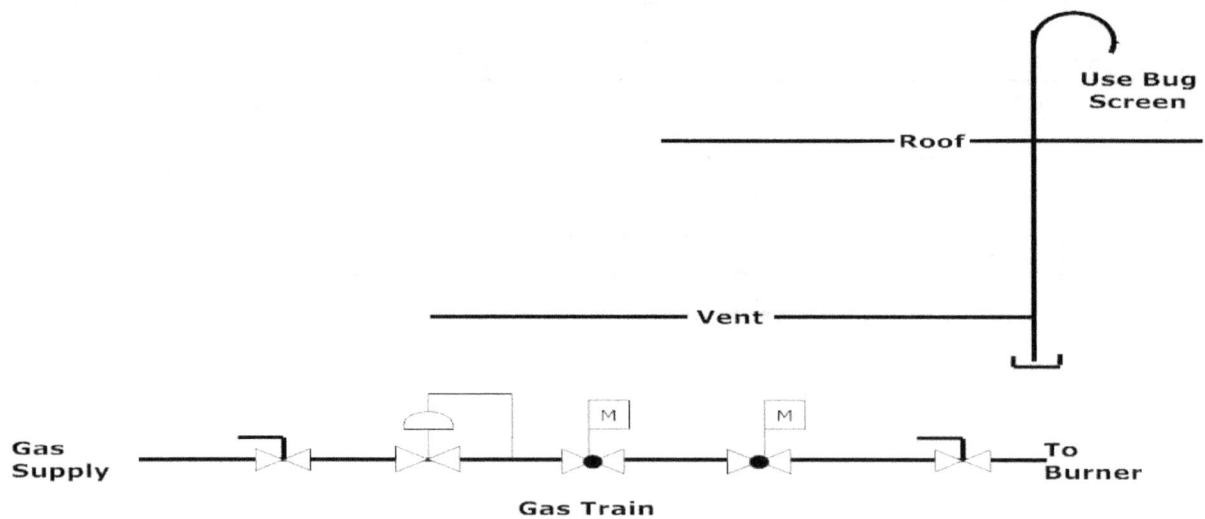

Some gas train components, such as regulators and gas pressure switches, have to be vented to the outside. A leaking 1/4" fitting could lose enough gas to fill a 10 foot by 10 foot x 10 foot high room with a combustible mixture within an hour. When combining common vents from gas train components, you need to see if the common pipe cross sectional area is large enough for all the components.

Let us assume that we will be venting two regulators with 3/4" vents and two gas pressure switches with 1/4" vents. We will get our cross sectional sizes from the chart below

3/4" regulators	0.533 X 2 = 1.066
1/4" gas pressures switches	0.104 X 2 = 0.208
	Total 1.274

We would need a pipe size of 1 1/4" to combine all these vents.

- Pipe runs over 30 feet horizontal are not recommended and should be increased one pipe size for each 30 feet run.
- Normally open vent valves cannot be combined with other vents or other appliances.

Vent terminations should be:

- 4 feet below, 1 foot above, and 3 feet horizontally from windows, doors, and gravity air intakes. 3 feet above any forced air inlet within 10 feet horizontally.
- Vents should have a screened 90 facing down.

Pipe Size	Inside Diameter	Inside Cross Sectional Area
1/8"	0.269	0.057
1/4"	0.364	0.104
1/2"	0.622	0.304
3/4"	0.824	0.533
1"	1.049	0.864
1 1/4"	1.38	1.495
1 1/2"	1.61	2.036
2"	1.61	2.036
2 1/2"	2.469	4.788
3"	3.068	7.393
4"	4.026	12.73
5"	5.047	20.004
6"	6.065	28.89

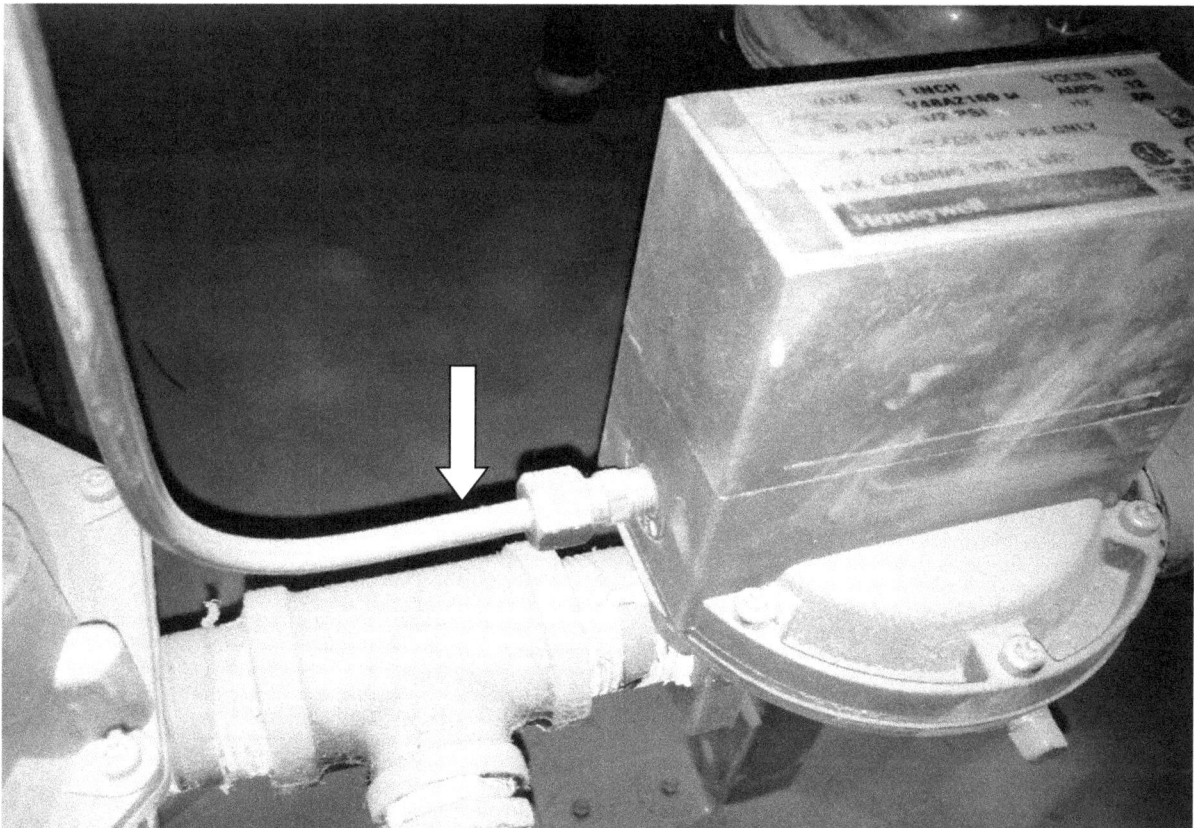

Velocity Calculation

Typical Hydronic Velocity = 2 -4.5 Feet per second in occupied areas. Slightly higher in unoccupied areas. Flows above 6 feet per second could erode copper. Flows below 2 could allow air to be trapped in the piping and cause air locked systems.

To find Fluid Velocity

$$Feet\ per\ Second\ \frac{0.408\ x\ GPM}{(Pipe\ Diameter\ Inches)^2}$$

$$Gallons\ per\ Minute\ GPM = (Pipe\ Diameter\ Inches)^2 \div FPM$$

Feet per second FPS to Miles per Hour MPH			
Feet per second	MPH	Feet per second	MPH
1	0.68	9	6.13
2	1.37	10	6.82
3	2.05	11	7.50
4	2.73	12	8.18
5	3.40	13	8.87
6	4.09	14	9.55
7	4.77	15	10.23
8	5.45	16	10.91

Pipe GPM @ Various Velocities Feet per Minute					
FPM = Feet per Minute					
Pipe Size Inches	2 FPM	3 FPM	4 FPM	5 FPM	6 FPM
Gallons per Minute					
1/2"	2	4	5	12	22
3/4"	4	6	8	21	38
1"	7	10	14	35	62
1 1/4"	12	18	24	60	108
1 1/2"	16	18	33	81	147
2"	27	24	54	135	242
3"	59	89	118	296	533
4"	102	153	204	510	918
6"	231	347	463	1,157	2,082
8"	400	600	800	2,000	3,599
10"	631	946	1,261	3,153	5,675
12"	895	1,343	1,791	4,476	8,058
14"	1,083	1,624	2,165	5,413	9,744
16"	1,413	2,120	2,825	7,065	12,717
18"	1,789	2,684	3,579	8,947	16,104
20"	2,222	3,333	4,444	11,110	19,998
24"	3,216	4,824	6,432	16,080	28,945

Convert Flow (GPM) to Velocity (FPS)

GPM	Pipe Diameter in Inches					
	2	4	6	8	10	12
	FEET PER SECOND					
5	0.51	0.13	0.06	0.03	0.02	0.01
10	1.02	0.26	0.11	0.06	0.04	0.03
15	1.53	0.38	0.17	0.10	0.06	0.04
20	2.04	0.51	0.23	0.13	0.08	0.06
30	3.06	0.77	0.34	0.19	0.12	0.09
40	4.08	1.02	0.45	0.26	0.16	0.11
50	5.10	1.28	0.57	0.32	0.20	0.14
60	6.12	1.53	0.68	0.38	0.24	0.17
70	7.14	1.79	0.79	0.45	0.29	0.20
80	8.16	2.04	0.91	0.51	0.33	0.23
90	9.18	2.30	1.02	0.57	0.37	0.26
100	10.20	2.55	1.13	0.64	0.41	0.28
150	15.30	3.83	1.70	0.96	0.61	0.43
200	20.40	5.10	2.27	1.28	0.82	0.57
250	25.50	6.38	2.83	1.59	1.02	0.71
300	30.60	7.65	3.40	1.91	1.22	0.85
400	40.80	10.20	4.53	2.55	1.63	1.13
500	51.00	12.75	5.67	3.19	2.04	1.42
600	61.27	15.30	6.80	3.83	2.45	1.70
700	71.40	17.85	7.93	4.46	2.86	1.98
800	81.60	20.40	9.07	5.10	3.26	2.27
900	91.80	22.95	10.20	5.74	3.67	2.55
1,000	102.00	25.50	11.33	6.38	4.08	2.83
1,500	153.00	38.25	17.00	9.56	6.12	4.25
2,000	204.00	51.00	22.67	12.75	8.16	5.67

To estimate Btuh

Btuh=500 x GPM x Δt
- Btuh = Btu/Hr
- GPM = Gallons per minute
- 500 = 8.33(Weight of one gallon of water) x 60 (minutes)
- Δt = Temperature difference F^0

One gpm will deliver approximately 10,000 Btuh with a 20 degree F delta T

Maximum flow through a boiler is boiler output divided by the temperature rise* divided by 500.
*Please note most boilers require a 20-30 degree temperature rise. 20^0F

How many Btus @ various boiler temperature Delta T		
20^0F Delta T	25^0F Delta T	30^0F Delta T
3.45 GPM per Boiler HP	2.88 GPM per Boiler HP	2.30 GPM per Boiler HP
10,000 Btuh/GPM	12,500 Btuh/GPM	15,000 Btuh/GPM

	20^0F Rise	25^0F Rise	30^0F Rise
Btuh Output	Gallons per minute		
400,000	40	32	27
500,000	50	40	33
600,000	60	48	40
700,000	70	56	47
800,000	80	64	53
900,000	90	72	60
1,000,000	100	80	67
1,100,000	110	88	73
1,200,000	120	96	80
1,300,000	130	104	87
1,400,000	140	112	93
1,500,000	150	120	100
1,600,000	160	128	107
1,700,000	170	136	113
1,800,000	180	144	120
1,900,000	190	152	127
2,000,000	200	160	133
3,000,000	300	240	200
5,000,000	500	400	333
10,000,000	1,000	800	667

Misc Boiler Information

Closed Vessel Boiling Point @ PSI @ Sea Level	
Boiling Temperature	Gauge Pressure
212^{0}F	0 PSI
240^{0}F	10 PSI
259^{0}F	20 PSI
274^{0}F	30 PSI
287^{0}F	40 PSI
298^{0}F	50 PSI
316^{0}F	70 PSI
331^{0}F	90 PSI

How Long Will It Last?
Source ASHRAE

Equipment	Years
Boilers	24-35
Burners	21
Boilers, Condensing	10-15*
Rooftop Units	15
Furnaces	18
Pumps, Base Mounted	20
Pumps Pipe Mounted	10
Pumps, Condensate	15
Condensate Piping	10-20
Steam Traps	*7 based upon US GSA*
* Based upon findings by Chartered Institute of Building Service Engineers	

Boiler Ratings

One boiler horsepower = 33,475 Btuh output

Boiler HP	Btu/Hr Output	Boiler HP	Btu/Hr Output
20	669,500	250	8,368,750
30	1,004,250	300	10,042,500
40	1,339,000	350	11,716,250
50	1,673,750	400	13,390,000
60	2,008,500	450	15,063,750
70	2,343,250	500	16,737,500
80	2,678,000	600	20,085,000
100	3,347,500	700	23,432,500
125	4,184,375	800	26,780,000
150	5,021,250	1,000	33,475,000
200	6,695,000		

Misc Boiler Information

1 Boiler HP =	**34,500 Btuh**
	34.5 Lbs Steam/ Hr from and at 212 degrees F
	34.5 Lb H2O/ Hr
	140 EDR
	about 10-11 ½ square feet of boiler heating surface
1 EDR hot water =	150 Btuh
1 Btu =	will raise 1 cubic feet of air 55 degrees
	will raise 55 cubic feet of air 1 degree F
	Amount of heat required to raise one pound of water one degree

Common Heating Calculations	
To Find?	**Perform this calculation**
Lbs Steam per Hour	Boiler HP x 34.5
Evaporation Rate GPM	Boiler HP x .069
MBTU per Hr Output (MBH)	Boiler HP x 33.4
Sq. Ft of Equivalent Direct Radiation (EDR)	Boiler HP x 139
Sq Ft Equivalent Direct Radiation (EDR)	BTU/Hr /240
Evaporation Rate GPM	EDR/1000 x 0.5
Evaporation rate GPM	Lbs Steam Hr / 500
BTU	500 x GPM x Delta T
GPM	BTU / 500 x Delta T
Delta T	BTU /500 x GPM
Calculate Sensible Heat on Fan	1.08 x CFM x Temp Rise or Delta T
Calculate Sensible Heat	500 x GPM x Temp Rise or delta T
Average Building Heat Loss	25-40 BTUH per square foot

Types of Heat	
Sensible Heat	Sensible heat is any heat transfer that causes a change in temperature without causing a change of state. Sensible heat can be measured with a dry bulb thermometer. For example, when the temperature is increased over a heating coil, the temperature differential is sensible heat.
Latent Heat	Latent heat is the amount of heat required to cause a change of state. In a boiler system, this would be the amount of heat added to water to cause it to change from water to steam. It requires 970.4 Btus to raise 1 pound of water at 212 degrees f to 1pound of steam. The latent heat that is added to change water to steam is also given back when the steam condenses in the radiator or coil.
Total Heat	Total heat is the sum of the sensible and latent heat in an exchange process. It is sometimes called enthalpy.

Types of Heat Transfer

Conduction is the transfer of heat through a material or substance. It could even transfer to the adjoining material. An example of this is the heat that transfers from the frying pan on a stove to the handle of the pan. In our industry, it is like the heat that transfers along a pipe as it is soldered. Heat is also conducted through the ceilings, walls and floors of homes.

Convection is the transfer of heat by a liquid or gas (such as air). Circulatory air motion due to warmer air rising and cooler air falling is a common mechanism by which thermal energy is transferred. In our industry, it would be the heat that is circulated in a room heated with fin tube radiation. Convective heat loss also occurs through cracks and holes in the building and gaps and voids in ceilings, walls, and floors—and in the insulation.

Radiation heat transfer occurs between objects that are not touching. The most common example of this is the way the sun heats the earth. The sun warms the earth without warming the space between the sun and the earth. An example in our industry is the heat that you feel from a radiant heater or large cast iron radiator.

PIPING
Recommended Maximum Pipe Flow Rates based upon 20^0F Delta T

Copper Pipe		
Pipe Size	Maximum Flow GPM	Btuh
½"	1 1/2	15,000
¾"	4	40,000
1"	8	80,000
1 ¼"	14	140,000
1 ½"	22	220,000
2"	45	450,000
2 ½"	85	850,000
3"	130	1,300,000

PEX Piping		
Pipe Size	Maximum Flow GPM	Btuh
3/8"	1.2	12,000
1/2"	2	20,000
5/8"	4	40,000
3/4"	6	60,000
1"	9.5	95,000

Steel Pipe		
Pipe Size	Maximum Flow GPM	Btuh
½"	2	15,000
¾"	4	40,000
1"	8	80,000
1 ¼"	16	160,000
1 ½"	25	250,000
2"	50	500,000
2 ½"	80	800,000
3"	140	1,400,000
4"	300	3,000,000
6"	850	8,500,000
8"	1,800	18,000,000
10"	3,200	32,000,000
12'	5,000	50,000,000

Number of smaller pipes that equal One larger pipe

Pipe Size	1/2	3/4	1	1 1/4	1 1/2	2	2 1/2	4	6
1/2	1.00	2.27	4.88	10.00	15.80	31.70	52.90	205	620
3/4		1.00	2.05	4.30	6.97	14.00	23.30	90	273
1			1.00	2.25	3.45	6.82	11.40	44	133
1 1/4				1.00	1.50	3.10	5.25	19	68
1 1/2					1.00	2.00	3.34	13	39
2						1.00	1.67	6.50	19.60
2 1/2							1.00	3.87	11.70
4								1.00	3.02
6									1.00

Standard Nipples & Pipe Sizing Schedule 40					
Pipe Size	Outside Diameter (O.D.)	Circumference	Pipe Size	Outside Diameter (O.D.)	Circumference
1/8"	0.405"	1.272"	2 1/2"	2.875"	9.032"
¼ "	0.540"	1.696"	3"	3.500"	10.995"
3/8"	0.675"	2.121"	4"	4.500"	14.137"
½"	0.840'	2.639"	5"	5.563"	17.476"
¾"	1.050"	3.299"	6"	6.625"	20.812"
1"	1.315"	4.131"	8"	8.625"	27.095"
1 ¼"	1.660"	5.215"	10"	10.750"	33.771"
1 ½"	1.900"	5.969"	12"	12.750"	40.054"
2"	2.375"	7.461"			

Standard Copper Tubing Type K,L,M					
Pipe Size	Outside Diameter (O.D.)	Circumference	Pipe Size	Outside Diameter (O.D.)	Circumference
½"	0.625"	1.964"	3"	3.125"	9.817"
¾"	0.875"	2.749"	4"	4.125"	12.959"
1"	1.125"	3.534"	6"	6.127"	12.248"
1 ¼"	1.375"	4.319"	8"	8.125"	25.525"
1 ½"	1.625"	5.105"	10"	10.125"	31.808"
2"	2.125"	6.675"	12"	12.750"	40.054"
2 1/2"	2.625"	8.246"			

PIPING RESISTANCE

Typical Friction head for pipe is 4.2' per 100 feet @500 Milinches restriction per foot

Frictional Allowance for Fittings In Feet of Pipe					
Length to be added in feet					
Size of Fittings Inches	90° Ell	Side Outlet Tee	Gate Valve	Globe Valve	Angle valve
1/2"	1.3	3	0.3	14	7
3/4'	1.8	4	0.4	18	10
1"	2.2	5	0.5	23	12
1 1/4"	3.0	6	0.6	29	15
1 1/2"	3.5	7	0.8	34	18
2"	4.3	8	1.0	46	22
2 1/2"	5.0	11	1.1	54	27
4"	9	18	1.9	92	45
6"	13	27	2.8	136	67
8"	17	35	3.7	180	92
10"	21	45	4.6	230	112

PIPING EXPANSION

Calculating the linear expansion of pipe carrying steam or hot water. If you would like to calculate the expansion or lengthening of a pipe when it has steam or hot water inside, try this formula

$$E = Constant * Temperature\ rise$$

E = Expansion in inches per 100 feet of pipe   F = Starting temperature
T = Final temperature   C = Constant

Constant = Coefficient of expansion per 100 Ft pipe

Metal	Constant
Steel	0.00804
Wrought Iron	0.00816
Cast Iron	0.00780
Copper or Brass	0.01140

For Example

What is the expansion of 100 feet of copper tubing that will heat water from 50^0F to 180^0F?

$$E = Constant * Temperature\ rise$$

$$E = 0.01140 * (180 - 50)Temp\ rise$$

$$1.482" = 0.01140 * 130$$

The expansion is 1.482" for every 100 feet of copper tubing.

What is the expansion of 100 feet of steel pipe that will heat water from 50^0F to 180^0F?

$$E = Constant * Temperature\ rise$$

$$E = 0.00804 * (180 - 50)Temp\ rise$$

$$1.045" = 0.00804 * 130$$

The expansion is 1.045" for every 100 feet of steel pipe.

Copper expands at 39% greater rate than steel

Thermal Expansion of Piping Material in inches per 100 feet from 32 Deg F

Temperature Rise Deg F	Carbon & Carbon Moly Steel	Cast Iron	Copper
	Inches	Inches	Inches
32^{0}F - 32^{0}F	0	0	0
32^{0}F - 100^{0}F	0.5	0.5	0.8
32^{0}F - 150^{0}F	0.8	0.8	1.4
32^{0}F - 200^{0}F	1.2	1.2	2.0
32^{0}F - 250^{0}F	1.7	1.5	2.7
32^{0}F - 300^{0}F	2.0	1.9	3.3
32^{0}F - 350^{0}F	2.5	2.3	4.0
32^{0}F - 400^{0}F	2.9	2.7	4.7
32^{0}F - 500^{0}F	3.8	3.5	6.0
32^{0}F - 600^{0}F	4.8	4.4	7.4
32^{0}F - 700^{0}F	5.9	5.3	9.0

Heat Loss from Pipe

Heat Losses from Uninsulated Horizontal Steam Pipe		
Btu per hour per linear foot at 70 Deg E room temperature		
Nom Pipe Size	Hot Water (180 Deg F)	Steam 5 PSIG
½	60	96
¾	73	118
1	90	144
1 ¼	112	170
1 ½	126	202
2	155	248
2 ½	185	296
3	221	355
4	279	448

Heat Losses from Uninsulated Horizontal Copper Pipe	
Btu per hour per linear foot at 70 Deg E room temperature	
Nom Pipe Size	Hot Water (180 Deg F)
½	33
¾	45
1	55
1 ¼	66
1 ½	77
2	97
2 ½	117
3	136
4	174

Avg. Loss from Insulated pipe Btu/ Linear foot @ 70 Deg F		
Pipe Size Inches	**Insulation Thickness**	**175⁰F**
½	1"	0.150
¾	1"	0.172
1	1"	0.195
	1 ½"	0.165
1 ¼"	1"	0.250
	1 ½"	0.170
1 ½"	1"	0.247
	1 ½"	0.205
2"	1"	0.290
	1 ½"	0.235
	2"	0.200
2 ½"	1"	0.330
	1 ½"	0.265
	2"	0.225
3"	1"	0.385
	1 ½"	0.305
	2"	0.257
4"	1"	0.470
	1 ½"	0.370
	2"	0.308

Sizing a Circulator

There are a couple short cuts to sizing a pump for a boiler. Most boilers are designed for a 20 to 30 degree rise through the boiler. To size a pump for a boiler and maintain a 20 degree rise through the boiler, divide the output of the boiler by 10,000 to get the proper GPM for a 20 degree rise.

If the boiler can handle a 30 degree rise, divide the output of the boiler by 15,000 to get the proper GPM.

Example: To see if the existing 40 GPM pump is large enough for the new boiler, let us look at the equipment. Our new boiler has a rated output of 800,000 with a design temperature rise of 20 degrees F. The existing pump is 40 GPM. If we divide the boiler output by 10,000, we see that the boiler will require an 80 GPM pump. This is double the GPM of the existing pump. Our flow would be half and the temperature rise would be double, possibly ruining the new boiler. In this case, we would have to replace the pump. If our new boiler can handle a 30 degree rise, we could divide it by 15,000 and find that the boiler will require a 53 GPM pump. The existing pump is too small for this boiler.

If you would like to see how we arrived at the 10,000 or 15,000 number, the following is the formula:

$$GPM = \frac{\text{Rated output of boiler}}{8.33 * 60 * \triangle \,°F}$$

or

$$GPM = \frac{\text{Rated output of boiler}}{500 * \triangle \,°F}$$

$$500 = 8.33 * 60$$

GPM = Gallons per minute flow rate
8.33 = Weight of a gallon of water
60 = Converts the formula from hours to minutes aka Gallons per Minute GPM
$\triangle$ °F Temperature rise through boiler is usually about 20-30 degrees F.

The following is the actual formula for a 20 degree rise for the 800,000 Btuh boiler:

$$GPM = \frac{800,000 \; Btuh \; (Output \; of \; boiler)}{8.33 * 60 * Temperature \; rise \; through \; boiler}$$

$$GPM = \frac{800,000\ Btuh\ (Output\ of\ boiler)}{500*20\ Degree\ rise}$$

$$80\ GPM = \frac{800,000\ Btuh\ (Output\ of\ boiler)}{10,000}$$

The following is the actual formula for a 30 degree rise for the 800,000 Btuh boiler:

$$GPM = \frac{800,000\ Btuh\ (Output\ of\ boiler)}{8.33*60*Temperature\ rise\ through\ boiler}$$

$$GPM = \frac{800,000\ Btuh\ (Output\ of\ boiler)}{500*30\ Degree\ rise}$$

$$53\ GPM = \frac{800,000\ Btuh\ (Output\ of\ boiler)}{15,000}$$

Calculate pump head

1. Measure longest run in feet. Include both supply and return.
2. Multiply by 1.5 to calculate fittings and valves
3. Multiply by 0.04 (4' head for each 100' of pipe ensures quiet operation)

For example, 100 feet is longest run

100 x 1.5 x .04 = 6 feet of head

Water Capacities

Water Capacity Steel Pipe			
Sch 40	**US Gallons per Foot**	**Sch 40**	**US Gallons per Foot**
Pipe Size Inches	**Water Capacity/ ft**	**Pipe Size Inches**	**Water Capacity/Ft**
½"	0.016	3"	0.390
¾"	0.023	4"	0.690
1"	0.040	5"	1.100
1 ¼"	0.063	6"	1.500
1 ½ "	0.102	8"	2.599
2"	0.170	10"	4.096
2 ½"	0.275	12"	5.815

Water Capacity Copper Tubing			
	US Gallons per Foot		
Pipe Size	**Type K**	**Type L**	**Type M**
3/8"	0.006	0.007	0.008
½"	0.011	0.012	0.013
5/8"	0.017	0.017	
¾"	0.023	0.025	0.027
1"	0.040	0.043	0.045
1 ¼"	0.063	0.065	0.068
1 ½"	0.089	0.092	0.095
2"	0.159	0.161	0.165
2 ½"	0.242	0.248	0.254
3"	0.345	0.354	0.363
4"	.608	.571	.634

Average System Water Content US Gallons	
Heating Element	Estimated Volume
Cast Iron Radiation	
Radiator, Large Tube	0.114 gal/ sq foot
Radiator, Thin Tube	0.056 gal/ sq foot
Convectors	1.5 Gal/10,000 Btu/Hr @ 200°F
Baseboard	4.7 Gal/10,000 Btu/Hr @ 200°F
Radiation Non Ferrous	
Convectors	0.64 Gal/10,000 Btu/Hr @ 200°F
Baseboard ¾"	.37 Gal/10,000 Btu/Hr @ 200°F
Fan Coil / Unit Htr.	.2 Gal/10,000 Btu/Hr @ 180°F

WATER FORMULAS

Water Scalding Times

Temperature ^0F	1st Degree	2nd/3rd Degree
111.20	5 Hrs	7 Hrs
116.60	35 Mins	45 Mins
118.40	10 Mins	14 Mins
122.00	1 Minute	5 Mins
131.00	5 Secs	22 Secs
140.00	2 Secs	5 Secs
149.00	1 Sec	2 Secs
158.00	1 Sec	1 Sec

pH Scale

pH	Equivalent	If pH was measured with dollars
PH = 0	Battery Acid	-$100,000,000
PH = 1	Hydrochloric Acid	-$10,000,000
PH = 2	Lemon Juice, Vinegar	-$1,000,000
PH = 3	Grapefruit, Orange Juice	-$100,000
PH = 4	Acid Rain, Tomato Juice	-$10,000
PH = 5	Black Coffee	-$1,000
PH = 6	Urine, Saliva	-$100
PH = 7	"Pure" Water	$10
PH = 8	Sea Water	$100
PH = 9	Baking Soda	$1,000
PH = 10	Milk of Magnesia	$10,000
PH = 11	Ammonia Solution	$100,000
PH = 12	Soapy Water	$1,000,000
PH = 13	Bleaches, Oven Cleaner	$10,000,000
PH = 14	Liquid Drain Cleaner	$100,000,000

Acidic pH at 5-6, could cause corrosion. High alkaline between 8-10 causes scale deposits on hot transfer surfaces. The column on the far right shows the equivalent readings if pH was measured with dollars.

NOTE: Typical pH from condensing boiler flues are between 2.9 to 4.0.

Water Density			
Temperature	Temperature	Density	
^{0}F	^{0}C	Lb/ft^3	Lb/ gallon
32	0	63.41	8.48
39	3.89	63.42	8.48
50	10	63.40	8.48
68	20	63.31	8.46
86	30	63.15	8.44
122	50	62.67	8.38
140	60	62.35	8.34
158	70	62.01	8.29
176	80	61.63	8.24
194	90	61.22	8.18
212	100	60.78	8.13

Typical Water Calculations	
One Cubic Foot Water =	62.43 lbs
	7.48 gallons
	29.92 quarts
One pound of water =	27.72 cubic inches
One Gallon of Water =	0.1337 Cubic Feet
	4 Quarts
	8 Pints
	16 Cups

WATER

One cubic foot of water = 7.5 gallons
One cubic foot of water = 1,728 Cubic Inches
One cubic foot of water = 62.4 Pounds
One Pd of Water = 27.72 Cu Inches@65 Deg F
To estimate static pressure in system, multiply highest riser by 0.43 to get pressure at lowest point of system. Always add 4 pounds to get the right pressure for the building.
To estimate pump horsepower required - Horsepower = (GPM x Total head in feet) / 3960
To estimate flow rate of water through a pipe in gallons per minute - GPM = 0.0408 x (pipe diameter)^2 x (water velocity)
To estimate weight of water in a given section of pipe in pounds- Lbs of water = 0.34 x pipe length(feet) x (pipe diameter)^2
Maximum water velocity in pipes should be less than 6 feet per second @ 200^{0}F

Water Conversion Factors

US Gallons	X	8.34	=	Pounds
US Gallons	X	0.1338	=	Cubic Feet
US Gallons	X	231	=	Cubic Inches
Cu Inch water	X	0.03613	=	Pounds
Cu Inch water	X	0.004329	=	US Gallons
Cu Inch water	X	0.576384	=	Ounces
Pounds Water	X	27.72	=	Cu Inches
Pounds Water	X	0.12	=	US Gallons
PSIG	X	2.307	=	Height of water in feet

Water Pressure to Feet Head			
Pounds Per Sq Inch	Feet Head	Pounds Per Sq Inch	Feet Head
1	2.31	100	230.90
2	4.62	110	253.98
3	6.93	120	277.07
4	9.24	130	300.16
5	11.54	140	323.25
6	13.85	150	346.34
7	16.16	160	369.43
8	18.47	170	392.52
9	20.78	180	415.61
10	23.09	200	461.78
15	34.63	250	577.24
20	46.18	300	692.69
25	57.72	350	808.13
30	69.27	400	922.58
40	92.36	500	1,154.48
50	115.45	600	1,385.39
60	138.54	700	1,616.3
70	161.63	800	1,847.2
80	184.72	900	2,078.1
90	207.81	1,000	2,309.00

Feet Head to Water Pressure			
Feet Head	Pounds Per Sq Inch	Feet Head	Pounds Per Sq Inch
1	.43	100	43.31
2	.87	110	47.64
3	1.30	120	51.97
4	1.73	130	56.30
5	2.17	140	60.63
6	2.60	150	64.96
7	3.03	160	69.29
8	3.46	170	73.63
9	3.90	180	77.96
10	4.33	200	86.62
15	6.50	250	108.27
20	8.66	300	129.93
25	10.83	350	151.58
30	12.99	400	173.24
40	17.32	500	216.55
50	21.65	600	259.85
60	25.99	700	303.16
70	30.32	800	346.47
80	34.65	900	389.78
90	38.98	1,000	433.00

Hot Water System Makeup:

Minimum connection size shall be 10% of largest system pipe or 1", whichever is greater 20" pipe should equal a 2" connection.

167

Sizing an Compression tank

Recommended Sizing for Compression tanks

Nominal Capacity Gallons	Sq ft Radiation
18	350
21	450
24	650
30	900
35	900
35	1100

A rule of thumb for sizing an compression tank is One gallon for each 23 square feet of radiation or One gallon for each 3,500 Btu of radiation. If you are going to size an compression tank, several of the manufacturers have on line calculators that will help you. If you still wish to do a manual calculation, here are the following calculations:

Compression tank Sizing

Closed Tank	$Vt = Vs \dfrac{[(V2/V1) - 1] - 3\alpha\Delta t}{(P\alpha/P1) - (P\alpha/P2)}$
Diaphragm Tank	$Vt = Vs \dfrac{[(V2/V1) - 1] - 3\alpha\Delta t}{1 - (P1/P2)}$

Definitions:
Vt = Volume of compression tank in gallons
Vs = Volume of water in system in gallons
V1= Ground water temperature
V2= Design heating water temperature (180^0F)
$\triangle T = T_2 - T_1$ ^{0}F
T_1 = Lower system temperature, typically 40-50^{0}F at fill condition.
T_2 = Higher system design temperature, typically 180^0- 220^{0}F.
$P\alpha$ = Atmospheric pressure (14.7 Psia)
P_1 = System fill pressure Minimum System pressure (Psia)
P_2 = System operating pressure Maximum System pressure (Psia)
α = Linear Coefficient of expansion
      Steel 6.5 x 10^{-6}
      Copper 9.5 x 10^{-6}

When choosing a diaphragm tank, use the acceptance factor when choosing the size. The acceptance factor is the amount of space that is available in the tank.

To see how this formula works, let us look at a hypothetical building. Our building has a system volume of 2,000 gallons. The system will operate between 180^0F and 220^0F. The minimum pressure will be 10 pounds and the maximum pressure will be at 25 pounds. 14.7 has to be added to each pressure to get atmospheric pressure. For example, the low pressure will be 14.7 pounds plus 10 pounds $= 24.7$ pounds. The high pressure will be 14.7 pounds plus 25 pounds $= 39.7$ pounds. Our relief valve is set at 30 pounds. The system is steel pipe. The volume of the water is as follows:

$V_1 = 40^0$F $= 0.01602$ ft^2/lb (Ground water temperature)

$V_2 = 220^0$F $= 0.01677$ ft^2/lb (Design Temperature)

$\alpha =$ coefficient of thermal expansion for steel pipe is 6.5 x 10^{-6}

We will size a closed tank. Here is the formula again.

$$Vt = Vs \frac{[(^{V2}/_{V1}) - 1] - 3\alpha\Delta t}{(^{P\alpha}/_{P1}) - (^{P\alpha}/_{P2})}$$

$$Vt = 2000 \frac{[(^{0.01677}/_{0.01602}) - 1] - 3(6.5 \, x \, 0.000001)x \, 180)}{(^{14.7}/_{24.7}) - (^{14.7}/_{39.7})}$$

$$Vt = 2000 \frac{0.0468 - 0.00351}{0.595 - 0.370}$$

$$Vt = 2000 \frac{0.0433}{.225}$$

Vt = 385 Gallons

Estimated system water volume 35 Gallons per Boiler HP

Typical system fill pressure 10 Psi

Rule of Thumb for Compression tank Sizing

Steel Piping
Entering Pressure 10 pounds
Maximum Pressure 25 pounds
Entering Temperature 40^{0}F
Maximum Temperature 220^{0}F

Copper Piping
Based upon the following:
Entering Pressure 10 pounds
Maximum Pressure 25 pounds
Entering Temperature 40^{0}F
Maximum Temperature 220^{0}F

Steel Piping				Copper Piping		
System Capacity in Gallons	Closed Compression tank	Diaphragm Tank		System Capacity in Gallons	Closed Compression tank	Diaphragm Tank
200	39	23		200	37	22
300	58	34		300	56	33
400	77	46		400	74	44
500	96	57		500	93	55
600	116	69		600	111	66
700	135	80		700	130	77
800	154	92		800	148	88
900	173	103		900	167	99
1,000	193	115		1,000	185	110
1,100	212	126		1,100	204	121
1,200	231	138		1,200	222	132
1,300	250	149		1,300	241	143
1,400	270	160		1,400	260	154
1,500	289	172		1,500	278	165
1,600	308	183		1,600	297	177
1,700	327	195		1,700	315	188
1,800	347	206		1,800	334	199
1,900	366	218		1,900	352	210
2,000	385	229		2,000	371	221
2,500	481	287		2,500	463	276
3,000	578	344		3,000	556	331
3,500	674	401		3,500	649	386
4,000	770	458		4,000	742	441
4,500	867	516		4,500	834	496
5,000	963	573		5,000	927	552
6,000	1,156	688		6,000	1,112	662
7,000	1,348	802		7,000	1,298	772
8,000	1,541	917		8,000	1,483	883
9,000	1,733	1,032		9,000	1,668	993
10,000	1,926	1,146		10,000	1,854	1,103

How to Estimate Hydronic System Volume

In many instances, you will need to calculate system water volume. This is useful when estimating water treatment and or glycol requirements. The following are some ideas that may help you to do that. The most accurate method is to measure and note the actual pipe sizes in the hydronic loop. This could be done by consulting the building blueprints. This is the most accurate method. There are several other rules of thumb that are used in the industry. I have a list of these below. A rule of thumb is that a hot water loop will be about 2/3 the size of a chilled water loop.

Rules of Thumb to Estimate System Volume
Multiply steel compression tank volume by 5.
35 – 50 gallons per Boiler HP
Pump GPM x 4
Compression tank volume is 20% of system volume
Rated tonnage of system x 10 gallons

The Salt Test
A common method for estimating system volume is to use salt because it is easy to test for, very soluble, and inexpensive. The disadvantage to this type of test is that the system has to be flushed at the end of the test, wasting water and chemicals. If not, the high chloride levels can be corrosive to the system.

Salt Test Procedure
Fill The system with fresh water. Circulate and flush the system until the water is clear. Eliminate all sources of water loss such as bleed, overflow, etc.

Measure the chloride Cl concentration in the system and estimate the system volume.

Add one pound of Table Salt (Sodium Chloride) per 1,000 gallons of estimated volume. This can be added in the pot feeder. Verify that the salt mixes thoroughly.

Allow one hour for the salt to be mixed into the system.

Re-Measure the chloride concentration

Multiply the estimated gallons of water by 76 ppm. Divide this by the difference (increase) in chloride concentration.

The answer will be the actual system volume

Example
Estimated Volume   1,000 Gallons
Initial Chloride Test   100 ppm
Final Chloride Test   180 ppm

Calculation

$$\frac{1,000\ gals. \times 76\ ppm}{(180-100)\ ppm} = 950 \text{ Actual Gallons in Loop}$$

Flush system rapidly to return the chloride level to normal.

Fuel & Flue Information & Piping

Flue Information

Vent Categories				
	I	II	III	IV
Vent Pressure	Negative	Negative	Positive	Positive
Temperature	>275^0	<275^0	>275^0	<275^0
Efficiency	<84%	>84%	<84%	>84%
Gas tight	No	No	Yes	Yes

The Products of Combustion Produced When One Cubic Foot of Gas is Burned	
One Cubic Foot of Gas Burned Produces	8 Cubic feet of nitrogen
	2 Cubic feet of water vapor
	1 Cubic Foot of Nitrogen

Typical Vent Temperature Ranges		
Venting Material	Temperature Ratings	Fuel
AL29-4C Stainless	0 - 480^0 F	Gas
B and BW Vent	0 - 550^0 F	Gas
L Vent	0 - 1,000^0 F	Oil
Factory Built Chimney	500^0 - 2,200^0 F	Oil/Gas
Masonry Chimney	360^0 - 1,800^0 F	Oil/Gas
Verify with manufacturer		

Condensing & Ignition Temperature of Various Fuels

Fuel	Condensing Temperature	Ignition Temperature
Natural Gas	250 ^{0}F	1,163 ^{0}F
#2 Oil	275 ^{0}F	600 ^{0}F
#6 Oil	300 ^{0}F	765 ^{0}F
Coal	325 ^{0}F	850 ^{0}F
Wood	400 ^{0}F	540-1,100 ^{0}F

Minimum Flue Gas Temperatures for Category 1 Boilers

Fuel	Minimum Flue Temperature
Natural Gas	265^{0}F plus 1/2^{0}F for each foot of stack or breeching, including horizontal and vertical runs
#2 Fuel Oil	240^{0}F plus 1/2^{0}F for each foot of stack or breeching, including horizontal and vertical runs

Acid Rain and Stack Temperature

Fuel	Acid Dew Point Temperature	Minimum Allowable Stack Temperature
Natural Gas	150	250
#2 Fuel Oil	180	275

Fuel

Comparative Fuel Values
To get 1,000,000 Btu's you need the following

Fuel Source	1,000,000 Btu's
Natural Gas @ 1000 Btu/ cu ft	1,000 Cu ft
Coal @ 12,000 Btu/ lb	83.333 Lb
Propane @ 91,600 Btu/ gal	10.917 Gal
Gasoline @ 125,000 Btus/gal	8.000 Gal
Fuel Oil #2 @ 140,000 Btus/gal	7.194 Gal
Fuel Oil #6 @ 150,000 Btus/gal	6.666 Gal
Electricity @ 3,412 Btu/kWh	293.083 Kwh

Average Btu Content of Common Fuels

Fuel Type	Number of Btu/ Unit
Fuel Oil #2	140,000 / Gallon
Fuel Oil #6	150,000 / Gallon
Butane	3,200 Btu's / CF
	21,500 Btus/ pound
	102,400 / Gallon
Natural Gas	1,025,000/ 1,000 cubic feet
Propane	91,330/ gallon
Coal	28,000,000 per ton
Electricity	3,412/ KWH
Wood MIxed	14,000,000/ cord or 3,500 / pound
Kerosene	135,000 / gallon
Pellets	16,500,000/ton

Misc Fuel Information

Natural Gas	
1 Cu ft Natural Gas =	1,000 Btus
1 MCF =	1,000,000 Btus
	1 MMBTU
	1 MCF
	1,000 Cu Ft.
	10 CCF
	10 Therms
1 Dekatherm	1 MCF
	10 Therms
	1,000,000 Btus
1 Therm =	100,000 Btus or 100 MBTU
	0.1 MCF
	1 CCF
	100 Cu. Feet
1 CCF =	1,000 Cu Ft
	100 Therm
Propane	
1 gallon =	92,000 Btus
1 Cu Foot =	2,250 Btus
#2 Fuel Oil	
1 Gallon =	140,000 Btus

Gas Pipe Line Sizing

Steel Pipe	Pipe Length			
Size	10 Feet	20 Feet	40 feet	80 Feet
	Capacity in Cubic Feet per hour			
½"	120	85	60	42
¾"	272	192	136	96
1"	547	387	273	193
11/4"	1,200	849	600	424
1 ½"	1,860	1,316	930	658
2"	3,759	2,658	1,880	1,330
2 ½"	6,169	4,362	3,084	2,189
4"	23,479	16,602	11,740	8,301

Each cubic foot of gas roughly equals 1,000 Btuh

Boiler Gas Consumption

Btuh	Cu Feet/Hr	Cubic Feet/ Min	Cubic Feet/ Sec
20,000,000	20,000	333.33	5.56
19,000,000	19,000	316.67	5.28
18,000,000	18,000	300.00	5.00
17,000,000	17,000	283.33	4.72
16,000,000	16,000	266.67	4.44
15,000,000	15,000	250.00	4.17
14,000,000	14,000	233.33	3.89
13,000,000	13,000	216.67	3.61
12,000,000	12,000	200.00	3.33
11,000,000	11,000	183.33	3.06
10,000,000	10,000	166.67	2.78
9,000,000	9,000	150.00	2.50
8,000,000	8,000	133.33	2.22
7,000,000	7,000	116.67	1.94
6,000,000	6,000	100.00	1.67
5,000,000	5,000	83.33	1.39
4,000,000	4,000	66.67	1.11
3,000,000	3,000	50.00	0.83
2,000,000	2,000	33.33	0.56
1,000,000	1,000	16.67	0.28
Based upon 1,000 Btu's per cubic foot			

Corrugated Stainless Steel Tubing CSST Sizing
Rules of Thumb for EHD Sizing EHD = Equivalent Hydraulic Diameter

EHD	Pipe Size	EHD	Pipe Size
15	3/8"	37	1 1/4"
19	1/2"	46	1 1/2"
25	3/4"	62	2"
31	1"		
Please check with the manufacturer to verify their sizing			

Inlet Pressure		Pressure Drop		Specific Gravity	
Less than 2 psi		0.6"		0.60	
Tube Size	Length				
EHD	5	10	15	20	25
Capacity in Cu ft per Hour					
13	46	32	25	22	19
15	63	44	35	31	27
18	115	82	66	58	52
19	134	95	77	67	60
23	225	161	132	116	104
25	270	192	157	137	122
30	471	330	267	231	206
31	546	383	310	269	240
37	895	639	524	456	409
46	1792	1260	1030	888	793
48	2070	1470	1200	1050	936
60	3660	2930	2400	2080	1860
EHD	30	40	50	60	70
13	18	15	13	12	11
15	25	21	19	17	16
18	47	41	37	34	31
19	55	47	42	38	36
23	96	83	75	68	63
25	112	97	87	80	74
30	188	162	144	131	121
31	218	188	168	153	141
37	374	325	292	267	248
46	723	625	559	509	471
48	856	742	665	608	563
60	1520	1320	1180	1080	1000

Gas Pressure Comparison

Inches Hg	Ounces	PSI
0.1	0.05	0.003
0.2	0.12	0.007
0.4	0.23	0.01
0.6	0.35	0.02
0.8	0.46	0.028
1	0.58	0.036
2	1.15	0.072
3	1.73	0.108
4	2.31	0.144
5	2.89	0.18
6	3.46	0.216
7	4.04	0.252
8	4.62	0.288
9	5.20	0.324
10	5.78	0.36
11	6.35	0.396
12	6.93	0.432
13	7.51	0.468
14	8.09	0.504
15	8.67	0.54
16	9.24	0.576
17	9.82	0.612
18	10.4	0.648
19	10.98	0.684
20	11.56	0.72
21	12.13	0.756
22	12.71	0.792
23	13.29	0.828
24	13.87	0.864
25	14.45	0.9
26	15.02	0.936
27	15.60	0.972
28	16.18	1.008
29	16.76	1.044
30	17.34	1.08
31	17.91	1.16
32	18.49	1.152
33	19.07	1.188

Pressure Conversion Chart
Inches H^{2}O to PSI 28" W.C. = 1 psi

Inches H^{2}O	PSI		Inches H^{2}O	PSI
0.1	0.0036		15	0.5414
0.2	0.0072		16	0.5774
0.4	0.0144		17	0.6136
0.6	0.0216		18	0.6496
0.8	0.0289		19	0.6857
1	0.0361		20	0.7218
2	0.0722		21	0.7579
3	0.1083		22	0.7940
4	0.1444		23	0.8301
5	0.1804		24	0.8662
6	0.2165		25	0.9023
7	0.2526		26	0.9384
8	0.2887		27	0.9745
9	0.3248		28	1.010
10	0.3609		29	1.047
11	0.3970		30	1.083
12	0.4331		31	1.191
13	0.4692		32	1.155
14	0.5053		33	1.191

Orifice Capacities for Natural Gas

1,000 Btu per cubic foot
Manifold pressure 3 ½" Water Column

Wire Gauge Drill Size	Rate Cu Ft / Hr	Rate Btu/Hr
70	1.34	1,340
68	1.65	1,650
66	1.80	1,870
64	2.22	2,250
62	2.45	2,540
60	2.75	2,750
58	3.50	3,050
56	3.69	3,695
54	5.13	5,125
52	6.92	6,925
50	8.35	8,350
48	9.87	9,875
46	11.25	11,250
44	12.62	12,625
42	15.00	15,000
40	16.55	16,550
38	17.70	17,700
36	19.50	19,500
34	21.05	12,050
32	23.70	23,075
30	28.50	28,500
28	34.12	34,125
26	37.25	37,250
24	38.75	38,750
22	42.50	42,500
20	44.75	44,750

Common Heating Abbreviations

Symbols			G	
Δt	Delta T or Temperature Difference		GPH	Gallons per hour
#	Pound		GPM	Gallons per minute
"	Inches		H	
'	Foot		HDD	Heating degree days
A			Hg	Mercury
AC	Alternating current		HHWP	Heating hot water pump
AHU	Air handling unit		HHWR	Heating hot water return
AMB	Ambient		HHWS	Heating hot water supply
B			HPS	High pressure steam
BTU	British Thermal Unit		Hs	Sensible heat
BTUH	British Thermal Unit per Hour		HVAC	Heating Ventilating & Air Conditioning
C			HWP	Hot water pump
Cc	Cubic centimeter		HWR	Hot water return
CFM	Cubic feet per minute		HWS	Hot water supply
CFS	Cubic feet per second		Hx	Heat exchanger
CI	Cast Iron		I	
CL	Center Line		In	Inches
CPVC	Chlorinated poly vinyl chloride		K	
Cu ft	Cubic foot		Kwh	Kilowatt per Hour
Cu In	Cubic Inches		L	
D			LAT	Leaving air temperature
DC	Direct current		Lb	Pound
DD	Degree day		LH	Latent heat
DEG	Degree		LL	Low Limit
Deg F or ^{0}F	Degree Fahrenheit		LPS	Low pressure steam
Deg C or ^{0}C	Degree Celsius		LRA	Locked rotor amps
DHW	Domestic Hot water		LWCO	Low water cutoff
Diam	Diameter		M	
E			MPT	Male pipe thread
EAT	Entering air temperature		N	
F			NC	Normally closed
Fp	Freezing Point		NO	Normally open
Fpm	Feet per minute		O	
			Oz	Ounce
			P	
Fps	Feet per second		PRV	Pressure reducing valve
Ft	Foot		Psi	Pounds per square inch
P			T	

Psia	Pounds per square inch, absolute		TDH	Total dynamic head
Psig	Pounds per square inch, gauge		TEMP	Temperature
PVC	Poly vinyl chloride		TH	Thermometer
R			**V**	
Rpm	Revolutions per minute		V	Volt
Rps	Revolutions per second		**W**	
RV	Relief valve		W	Watt
S			WC	Water column
Sec	Second		Whr	Watt Hour
Sp gr	Specific gravity		Wmin	Watt minute
Sp ht	Specific heat			
Sq ft	Square foot			
Sq in	Square inch			

1 W =	**0.00134 hp**
	3.414 Btu
	0.0035 lb of water evaporated per hour
	44.236 foot-pounds minute
	2,654.16 foot-pounds hour
1 kW =	1,000 W
	1.34 hp
	3.53 lbs water evaporated per hr from and at 212 degrees F
	0.955 Btu's
	57.3 BTU per minute
	3,413 Btuh per hour
1 HP =	746 W
	0.746 kW
	33,000 ft-lb per minute
	550 ft-lb per second
	33,475 Btuh
	34.5 Lbs Steam/hr from and at 212Degrees F
	42.746 Btu/min
	2,564.76 Btuh
1 Kwh =	1,000 W/Hr
	1.34 hp/hr
	3,600,000 joules
	3.53 lbs water evaporated per hr from and at 212 degrees F
	22.75 lbs water raised from 62 degrees F to 212 degrees F
1 Btu =	17.452 W/Min
	0.2909W/Hr
	17.452 watts per minute
	0.2909 watts hour

ELECTRICAL

Amp of Copper Wire Types
Single wire in open air

Wire Size AWG	TH UF	FEPW, RH, RHW, TWH, THWN, ZW, THHW, XHHW	USE-2, XHH, XHHW, TBS, SA, SIS, FEP, MI, RHW-2, THHN, ZW-2, THWN-2, FEPB, RHH, THHW, THW-2
0000	300	360	405
000	260	310	350
00	225	265	300
0	195	230	260
1	165	195	220
2	140	170	190
3	120	145	165
4	105	125	140
6	80	95	105
8	60	70	80
10	40	50	55
12	30	35	40
14	25	30	35
16	-	-	24
18	-	-	18

Up to 86-degree ambient temperature

Amp of Copper Wire Types
Three wires in cable

Wire Size AWG	TH UF	FEPW, RH, RHW, TWH, THWN, ZW, THHW, XHHW	USE-2, XHH, XHHW, TBS, SA, SIS, FEP, MI, RHW-2, THHN, ZW-2, THWN-2, FEPB, RHH, THHW, THW-2
0000	195	230	260
000	165	200	225
00	145	175	195
0	125	150	170
1	110	130	150
2	95	115	130
3	85	100	110
4	70	85	95
6	55	65	75
8	40	50	55
10	30	35	40
12	25	25	30
14	20	20	25
16	-	-	18
18	-	-	14

Up to 86-degree ambient temperature

How to calculate Electrical Phase Imbalance

% of Voltage Imbalance = Maximum deviation from average ÷ average x 100

Take average of all 3 readings between all legs to get average. Most motors limit imbalance to 2%

% Imbalance	% Motor Winding Temperature Increase	% Imbalance	% Motor Winding Temperature Increase
2%	8%	5%	50%
3%	18%	6%	72%
4%	32%	7%	98%

Standard 24 Volt Thermostat Connections		
Terminal	Usage	Normal Colors
R or V	24 VAC power	Red
Rh or 4	24 VAC Heating Power	Red
Rc	24 VAC Cooling Power	Red
C	24 VAC Common	Black
Y	1st Stage Cooling	Yellow
Y2	2nd Stage Cooling	Blue or Orange
W	1st Stage Heat	White
W2	2nd Stage Heat	No Standard Color
G	Fan	Green

Ohm's Law

Volts =	$\sqrt{Watts \times Ohms}$	**Amperes =**	$\dfrac{Volts}{Ohms}$
	$\dfrac{Watts}{Amperes}$		$\dfrac{Watts}{Volts}$
	Amperes x Ohms		$\sqrt{\dfrac{Watts}{Ohms}}$
Ohms =	$\dfrac{Volts}{Amperes}$	**Watts =**	Volts x Amperes
	$Watts/Amperes^2$		$Amperes^2$ x Ohms
	$\dfrac{Volts^2}{Watts}$		$\dfrac{Volts^2}{Ohms}$

Electrical Equivalents Formulas	
Watt=	44.236 foot-pounds minute
	2,654.16 foot-pounds hour
	0.00134 hp
	3.414 Btu
	0.0035 lb of water evaporated per hour
	44.236 foot-pounds minute
	2,654.16 foot-pounds hour
Kilowatt=	44,235 foot-pounds minute
	1.34 H.P.
	0.955 BTU per second
	57.3 Btu per minute
	3,438 Btu per hour
	1,000 W
	1.34 hp
	3.53 lbs water evaporated per hr from and at 212 degrees F
	0.955 Btu's
	57.3 BTU per minute
	3,413 Btuh per hour
1 H.P.=	33,000 foot-pounds minute
	746 watts
	42.746 Btu per minute
	2,564.76Btu per hour
1 Btu=	772 ft lbs
	17.452 watts per minute
	0.2909 watts hour

Electrical Formulas

Calculate Motor HP from Meter Readings	
DC Motors	$HP = \dfrac{V * A * Ef}{746}$
Single Phase AC Motors	$HP = \dfrac{V * A * Ef * PF}{746}$
Three Phase AC Motors	$HP = \dfrac{V * A * Ef * PF * 1.73}{746}$
V= Volts	
Ef= Motor Efficiency	
A= Amps	
PF= Power factor	

Full Load Amperes of Single Phase Motors			
HP	RPM	115V	230V
1/8	1725	2.8	1.4
	1140	3.4	1.7
	860	4.0	2.0
1/4	1725	4.6	2.3
	1140	6.15	3.07
	860	7.5	3.75
1/3	1725	5.2	2.6
	1140	6.25	3.13
	860	7.35	3.67
1/2	1725	7.4	3.7
	1140	9.15	4.57
	860	12.8	6.4
3/4	1725	10.2	5.1
	1140	12.5	6.25
	860	15.1	7.55
1	1725	13.0	6.5
	1140	15.1	7.55
	860	15.9	7.95

Full Load Amperes of Three Phase Motors			
HP	RPM	115V	230V
1/4	1725	0.95	0.48
	1140	1.4	0.7
	860	1.6	0.8
1/3	1725	1.19	0.6
	1140	1.59	0.8
	860	1.8	0.9
1/2	1725	1.72	0.86
	1140	2.15	1.08
	860	2.38	1.19
3/4	1725	2.46	1.23
	1140	2.92	1.46
	860	3.26	1.63
1	1725	3.19	1.6
	1140	3.7	1.85
	860	4.12	2.06
1 1/2	1725	4.61	2.31
	1140	5.18	2.59
	860	5.75	2.88
2	1725	5.98	2.99
	1140	6.5	3.25
	860	7.28	3.64
3	1725	8.70	4.35
	1140	9.25	4.62
	860	10.3	5.15
5	1725	14.0	7.0
	1140	14.6	7.3
	860	16.2	8.1
7 1/2	1725	20.3	10.2
	1140	20.9	10.5
	860	23.0	11.5

Typical Flame Safeguard Signals Siemens	
Gas Burner Controls	
Model	**Flame Signal**
LFL with UV sensor QRA **Minimum 70 uA DC**	100-450 µA DC Typical
LFL with Flame Rod **Minimum 6 uA DC**	20-100 uA DC
Oil Burner Controls	
Model	**Flame Signal**
LAL1 with photoresistive detector, QRB1	95-160 µA DC
LAL1 with blue-flame detector, QRC1	80-130µA DC
LAL2/LAL3 with photoresistive detector, QRB1	8-35 µA DC
LAL2/LAl3 with selenium photocell detector, RAR	6.5-30 µA DC
LAL4 with photoresistive detector, QRB1	95-160 µA DC
LAL4 with blue-flame detector, QRC1	80-130 µA DC

Typical Flame Safeguard Signals Fireye & Honeywell	
Fireye	
Model	Average Flame Signal
UVM	4.0-5.5 VDC
TFM	14-17 VDC
D-10/20/30	16-25 VDC
E-100/ 110	20-80 VDC
E-100/E110 with EPD Programmer	4-10 VDC
M Series II	4-6 VDC
Micro M Series	4-10 VDC
Micro M Series w Display	20-80 VDC
Honeywell	
Model	Average Flame Signal
RA890	2-6 µA DC
R4795	2-6 µA DC
R7795	2-6 µA DC
R4140	2-6 µA DC
R4150	2-6 µA DC
BC7000	2-6 µA DC
RM7890	1.25-5 VDC
RM7895	1.25 VDC
RM7840	1.25 VDC
RM7800	1.25 VDC

Flame Safeguard Definitions	
µA = Micro Amps	VDC = Volts DC
1 Micro Amp or µA = 0.000001 Amps	
1 Amp = 1,000,000 Micro Amps	
1 Amp = 1,000 Milliamps or mA	

Estimate Storage Tank Capacity in Gallons
Rectangular Tank
Sizing a Storage Tank

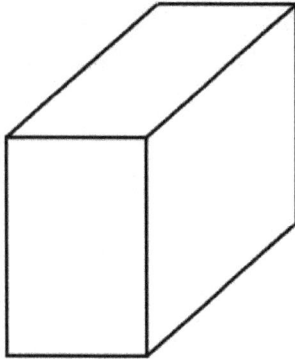

Rectangular Storage Tanks
Length" x Width" x Height" divided by 231 = Gallons
Measure tank length, width and height in inches
Multiply Length x Width x Height
Divide that total by 231 = Gallons
The following are some tank sizes that are already calculated.

	Rectangular Tank Height 12"			
	US Gallons			
	Width			
Length	12	18	24	36
12	7	11	15	22
18	11	17	22	34
24	15	22	30	45
30	19	28	37	56
36	22	34	45	67
42	26	39	52	79
48	30	45	60	90
60	37	56	75	112

Rectangular Tank Height 18"				
	US Gallons			
	Width			
Length	12	18	24	36
12	11	17	22	34
18	17	25	34	50
24	22	34	45	67
30	28	42	56	84
36	34	50	67	101
42	39	59	79	118
48	45	67	90	135
60	56	84	112	168

Rectangular Tank Height 24"				
	US Gallons			
	Width			
Length	12	18	24	36
12	15	22	30	45
18	22	34	45	67
24	30	45	60	90
30	37	56	75	112
36	45	67	90	135
42	52	79	105	157
48	60	90	120	180
60	75	112	150	224

Rectangular Tank Height 36"				
	US Gallons			
	Width			
Length	12	18	24	36
12	22	34	45	67
18	34	50	67	101
24	45	67	90	135
30	56	84	112	168
36	67	101	135	202
42	79	118	157	236
48	90	135	180	269
60	112	168	224	337

Rectangular Tank Height 48"				
	US Gallons			
	Width			
Length	12	18	24	36
12	30	45	60	90
18	45	67	90	135
24	60	90	150	269
30	75	112	150	224
36	90	135	180	269
42	105	157	209	314
48	120	180	239	359
60	150	224	299	449

Estimate Storage Tank Capacity in Gallons
Round Tanks

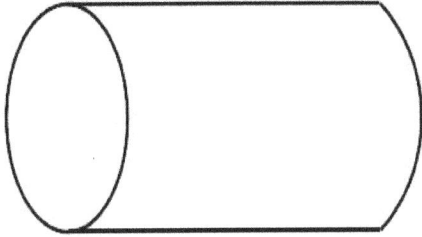

Multiply ½ Tank Diameter by itself
Multiply that by 3.146 x Length of Tank in inches
Divide by 231 = Gallons of water
For example, you have a 24" diameter tank that is 36" long
½ of 24" = 12" x 12" = 144
144 x 3.146 x 36 = 16,308
16,308 divided by 231 = 70.59 Gallons

Length (feet)	Circular Tank Inside Diameter (Inches)				
	US Gallons				
	18	24	30	36	42
1	1.1	1.96	3.06	4.41	5.99
2	26	47	73	105	144
2.5	33	59	91	131	180
3	40	71	100	158	216
3.5	46	83	129	184	252
4	53	95	147	210	288
4.5	59	107	165	238	324
5	66	119	181	264	360
5.5	73	130	201	290	396
6	79	141	219	315	432
6.5	88	155	236	340	468
7	92	165	255	368	504
7.5	99	179	278	396	540
8	106	190	291	423	576
9	119	212	330	476	648
10	132	236	366	529	720
12	157	282	440	634	864
14	185	329	514	740	1008

Circular Tank Inside Diameter (Inches)					
US Gallons					
Length (feet)	48	54	60	66	72
1	7.83	9.91	12.24	14.41	17.62
2	188	238	294	356	423
2.5	235	298	367	445	530
3	282	357	440	534	635
3.5	329	416	513	623	740
4	376	475	586	712	846
4.5	423	534	660	800	852
5	470	596	734	899	1057
5.5	517	655	808	978	1163
6	564	714	880	1066	1268
6.5	611	770	954	1156	1374
7	658	832	1028	1244	1480
7.5	705	889	1101	1355	1586
8	752	949	1175	1424	1691
9	846	1071	1322	1599	1903
10	940	1189	1463	1780	2114
12	1128	1428	1762	2133	2537
14	1316	1666	2056	2490	2960

MISC

Pressure Unit Conversions

Known	Desired Pressure Unit			
	Pounds Per sq In.	Ounces Per Sq In.	Inches of Water	Feet of Water
Centimeters of Water	0.0981	0.227	0.384	0.0328
Feet of water	0.433	6.94	12.0	0.883
Inches Mercury	0.491	7.86	13.6	1.13
Inches Water	0.0361	0.578	-------	0.0833
Ounces per Square Inch	0.0625	---------	1.73	0.144
Pounds per Sq Inch	----------	16.0	27.7	2.31

Common Fraction to Decimal to Millimeters

Fraction	Decimal	Millimeters
1/16	0.0625	1.587
1/8	0.125	3.175
3/16	0.1875	4.762
1/4	0.250	6.350
5/16	0.3125	7.937
3/8	0.375	9.525
7/16	0.4375	11.113
1/2	0.50	12.700
9/16	0.5625	0.5625
5/8	0.625	15.875
11/16	0.6875	17.462
3/4	0.750	19.050
13/16	0.8125	20.637
7/8	0.875	22.225
15/16	0.9375	23.812
1	1.00	25.400

Metric

Metric Liquid	
Metric	U.S.
3.7854 L	1 Gallon
0.946 L	1 Quart
0.473 L	1 Pint
1 L	0.264 Gallons
1 L	33.814 Ounces
29.576 ml	1 Fluid Ounce
236.584 ml	1 Cup
Metric Length	
Metric	U.S.
1m	39.37 inches
1 m	3.28 feet
1 m	1.094 yards
1 m	0.0016 miles
1.609 km	1 mile
25.4 mm	1 inch
2.54 cm	1 inch
304.8 mm	1 foot
1 mm	0.03937 inches
1 cm	0.3937 inches
1 dm	3.937 inches
Metric Pressure	
6.8947 kPa	1 pound per sq in (psi)
9.794 kPa	1m column of water
1 kPa	10.2 cm of water
1.3332 kPa	1 cm column of water
3.3864 kPa	1 inch of mercury (in Hg)
8 kPa	6 cm of mercury
Metric Conversions	
KJ/Hr =	Btu/h x 1.055
CMM =	CFM x 0.02832
LPM =	GPM x 3.785
Kj/Lb.=	Btu/Lb x 2.326
Meters =	Feet x 0.3048
Sq Meters =	Sq Feet x 0.0929
Cu. Meters =	Cu. Feet x 0.02832
Kg =	Pounds x 0.4536
Kg/Cu. Meter =	Pounds. Cu Feet x 16.017
Cu. Meters/ Kg =	Cu. Ft/ Pound x 0.0624

Fahrenheit to Celsius

F	C		F	C
1	-17.2		41	5.0
2	-16.7		42	5.6
3	-16.1		43	6.1
4	-15.6		44	6.7
5	-15.0		45	7.2
6	-14.4		46	7.8
7	-13.9		47	8.3
8	-13.3		48	8.9
9	-12.8		49	9.4
10	-12.1		50	10.0
11	-11.7		51	10.6
12	-11.1		52	11.1
13	-10.6		53	11.7
14	-10.0		54	12.2
15	-9.4		55	12.8
16	-8.9		56	13.3
17	-8.3		57	13.9
18	-7.8		58	14.4
19	-7.2		59	15.0
20	-6.7		60	15.6
21	-6.1		61	16.1
22	-5.6		62	16.7
23	-5.0		63	17.2
24	-4.4		64	17.8
25	-3.9		65	18.3
26	-3.3		66	18.9
27	-2.8		67	19.4
28	-2.2		68	20.0
29	-1.7		69	20.6
30	-1.1		70	21.1
31	-0.6		71	21.7
32	0		72	22.2
33	0.6		73	22.8
34	1.1		74	23.3
35	1.7		75	23.9
36	2.2		76	24.4
37	2.8		77	25.0
38	3.3		78	25.6
39	3.9		79	26.1
40	4.4		80	26.7

Fahrenheit to Celsius

F	C		F	C
81	27.2		121	49.4
82	27.8		122	50
83	28.3		123	50.5
84	28.9		124	51.1
85	29.4		125	51.7
86	30.0		126	52.2
87	30.6		127	52.8
88	31.1		128	53.3
89	31.7		129	53.9
90	32.2		130	54.4
91	32.8		131	55
92	33.3		132	55.6
93	33.9		133	56.1
94	34.4		134	56.7
95	35.0		135	57.2
96	35.6		136	57.8
97	36.1		137	58.3
98	36.7		138	58.9
99	37.2		139	59.4
100	37.8		140	60
101	38.3		141	60.6
102	38.8		142	61.1
103	39.4		143	61.7
104	40		144	62.2
105	40.6		145	62.8
106	41.1		146	63.3
107	41.6		147	63.9
108	42.2		148	64.4
109	42.8		149	65
110	43.3		150	65.6
111	43.8		151	66.1
112	44.4		152	66.7
113	45		153	67.2
114	45.6		154	67.8
115	46.1		155	68.3
116	46.6		156	68.9
117	47.2		157	69.4
118	47.7		158	70
119	48.3		159	70.56
120	48.9		160	71.1

Fahrenheit to Celsius

F	C		F	C
161	71.7		191	88.3
162	72.2		192	88.9
163	72.8		193	89.4
164	73.3		194	90
165	73.9		195	90.6
166	74.4		196	91.1
167	75		197	91.7
168	75.6		198	92.2
169	76.1		199	92.8
170	76.7		200	93.3
171	77.2		201	93.9
172	77.8		202	94.4
173	78.3		203	95
174	78.9		204	95.6
175	79.4		205	96.1
176	80		206	96.7
177	80.6		207	97.2
178	81.1		208	97.8
179	81.7		209	98.3
180	72.2		210	98.9
181	82.8		211	99.4
182	83.3		212	100
183	83.9		213	100.6
184	84.4		214	101.1
185	85		215	101.7
186	85.6		216	102.2
187	86.1		217	102.8
188	86.7		218	103.3
189	87.2		219	103.9
190	87.8		220	104.4

Misc.

	1,760 yards
1 Mile =	5,280 feet
	63,360 inches
	1.609 Km
1 Foot =	0.3048 M
	30.48 Cm
	304.8 mm
1 Inch =	25,400 microns
1 Gallon H2O =	8.333 Lbs
1 Lb =	16 oz
	7,000 grains
	0.4536 Kg
	1Lb = #
1 ton =	2,000 lbs
	907 Kg
1 lb steam =	1 lb H2O
1 square foot =	144 square inches
1 acre =	43,560 Sq Ft
	4,840 Sq Yds
	0.4047 Hectares
1 Sq Mile =	640 Acres
1 Sq Yd =	9 Sq Ft
	1,296 Sq Inches
1 Sq Foot =	Square yards x 9
1 Cu Yard =	27 Cu Ft
	46,656 cu inches
	1,616 pints
	807.9 quarts
	764.6 Liters
1 Cu foot =	1,728 cubic inches
1 Liter =	0.2642 Gallons
	1.057 quarts
	2.113 pints
1 gallon =	4 quarts
	8 pints
	3.785 liters
	0.13368 Cu Feet
	231 Cu Inches
1 Barrel Oil =	42 gallons
1 MPH =	5280 ft / hr
	88 ft/min

	1.467 ft/sec
	0.868 Knots per Hr
1 Knot =	1.1515 MPH
1 League =	3.0 Miles
Speed of Sound in Air =	1,128.5 ft/sec
	769.4 mph
14.7 psi =	33.95 ft H2O
	29.92 in Hg
	407.2 In wg
	2,116.8 Lbs/ Sq Ft
1 Psia =	Psig = 14.7
1 psi =	2.307 Ft H2O
	2.036 In Hg
	16 ounces
	27.7 In w.c.
1 ounce =	1.73 inches w.c
1 Ft H20 =	0.4335 psi
	62.43 lbs/ sq feet
Diameter of Circle =	Circumference x 0.3188
Circumference of Circle =	Diameter x 3.1416
Hours in a Year	8,760

Diameter to Circumference in Inches

Diameter	Circumference	Diameter	Circumference
12	37.70	28	87.96
14	43.98	30	94.25
16	50.27	32	100.53
18	56.55	34	106.81
20	62.83	36	113.10
22	69.12	38	119.38
24	75.40	40	125.66
26	81.68		

This page is left blank intentionally, unless there is a hidden message written in invisible ink that is only seen on full moon nights during October. You must decide that yourself. It is actually not really blank if I have this message on it so please ignore the above statement as it is mindless rambling.

Degree Day Definition

The daily mean temperature is obtained by adding together the maximum and minimum temperatures reported for the day and dividing the total by two. Each degree of mean temperature below 65 is counted as one heating degree-day. Thus, if the maximum temperature is 70^{0}F and minimum 52^{0}F, four heating degree-days would be produced. (70 + 52 = 122; 122 divided by 2 = 61; 65-61 = 4.) If the daily mean temperature is 65 degrees or higher, the heating degree-day total is zero.

Degree Days and Design Temperatures

ST	Station	Heating Degree Days	Heating Design Temp F
AL	Birmingham	2551	21
	Huntsville	3,070	16
	Mobile	1,560	29
	Montgomery	2,291	25
AK	Anchorage	10,864	-18
	Fairbanks	14,279	-47
	Juneau	9,075	1
	Nome	14,171	-27
AZ	Flagstaff	7,152	4
	Phoenix	1,765	34
	Tucson	1,800	32
	Yuma	974	39
AR	Fort Smith	3,292	17
	Little Rock	3,219	20
	Texarkana	2,533	23
CA	Fresno	2,611	30
	Long Beach	1,803	43
	Los Angeles	2,061	43
	Los Angeles	1,349	40
	Oakland	2,870	36
	Sacramento	2,502	32
	San Diego	1,458	44
	San Francisco	3,015	38
	San Francisco	3,001	40
CO	Alamosa	8,529	-16
CO	Colorado Springs	6,423	2
	Denver	6,283	1
	Grand Junction	5,641	7
CO	Pueblo	5,462	0
CT	Bridgeport	5,617	9

ST	Station	Heating Degree Days	Heating Design Temp F
	Hartford	6,235	7
	New Haven	5,897	7
DE	Wilmington	4,930	14
DC	Washington	4,224	17
FL	Daytona		
	Fort Myers	442	44
	Jacksonville	1,239	32
	Key West	108	57
	Miami	214	47
	Orlando	766	38
	Pensacola	1,463	29
	Tallahassee	1,485	30
	Tampa	683	40
	West Palm Beach	253	45
GA	Athens	2,929	22
	Atlanta	2,961	22
	Augusta	2,397	23
	Columbus	2,383	24
	Macon	2,136	25
	Rome	3,326	22
	Savannah	1,819	27
HI	Hilo	0	62
HI	Honolulu	0	63
ID	Boise	5,809	10
	Lewiston	5,542	6
	Pocatello	7,033	-1
IL	Chicago (Midway)	6,155	0
	Chicago (O'Hare)	6,639	-4
	Chicago	5,882	2
	Moline	6,408	-4
IL	Peoria	6,025	-4
	Rockford	6,830	-4
	Springfield	5,429	2
IN	Evansville	4,435	9
	Fort Wayne	6,205	1

State	City		
	Indianapolis	5,699	2
	South Bend	6,439	1
IA	Burlington	6,114	-3
	Des Moines	6,588	-5
	Dubuque	7,376	-7
	Sioux City	6,951	-7
	Waterloo	7,320	-10
KS	Dodge City	4,986	5
	Goodland	6,141	0
	Topeka	5,182	4
	Wichita	4,620	7
KY	Covington	5,265	6
	Lexington	4,683	8
	Louisville	4,660	10
LA	Alexandria	1,921	27
	Baton Rouge	1,560	29
	Lake Charles	1,459	31
	New Orleans	1,385	33
	Shreveport	2,184	25
ME	Caribou	9,767	-13
	Portland	7,511	-1
MD	Baltimore	4,654	13
	Baltimore	4,111	17
	Frederick	5,087	12
MA	Boston	5,634	9
	Pittsfield	7,578	-3
	Worcester	6,969	4
MI	Alpena	8,506	-6
MI	Detroit (city)	6,232	6
	Escanaba	8,481	-7
	Flint	7,377	1
	Grand Rapids	6,894	5
	Lansing	6,909	1
	Marquette	8,393	-8
	Muskegon	6,696	6
	Sault Ste. Marie	9,048	-8
MN	Duluth	10,000	-16
MN	Minneapolis	8,382	-12
	Rochester	8,295	-12
MS	Jackson	2,239	25
	Meridian	2,289	23
	Vicksburg	2,041	26
MO	Columbia	5,046	4
	Kansas City	4,711	6
	St. Joseph	5,484	2
	St. Louis	4,900	6

State	City		
	St. Louis	4,484	8
	Springfield	4,900	9
MT	Billings	7,049	-10
	Great Falls	7,750	-15
	Helena	8,129	-16
	Missoula	8,125	-6
NE	Grand Island	6,530	-3
	Lincoln	5,864	-2
	Norfolk	6,979	-4
	North Platte	6,684	-4
	Omaha	6,612	-3
	Scottsbluff	6,673	-3
NV	Elko	7,433	-2
	Ely	7,733	-4
	Las Vegas	2,709	28
	Reno	6,332	10
	Winnemucca	6,761	3
NH	Concord	7,383	-3
NJ	Atlantic City	4,812	13
	Newark	4,589	14
	Trenton	4,980	14
NM	Albuquerque	4,348	16
	Raton	6,228	1
	Roswell	3,793	18
NM	Silver City	3,705	10
NY	Albany	6,875	-1
	Albany	6,201	1
	Binghamton	7,286	1
	Buffalo	7,062	6
	NY (central park)	4,871	15
	NY (Kennedy)	5,219	15
	NY (LaGuardia)	4,811	15
	Rochester	6,748	5
NY	Schenectady	6,650	1
	Syracuse	6,756	2
NC	Charlotte	3,181	22
	Greensboro	3,805	18
	Raleigh	3,393	20
	Winston-Salem	3,595	20
ND	Bismarck	8,851	-19
	Devils Lake	9,901	-21
	Fargo	9,226	18
	Williston	9,243	-21
OH	Akron-Canton	6,037	6

	Cincinnati	4,410	6
	Cleveland	6,351	5
	Columbus	5,660	5
	Dayton	5,622	4
	Mansfield	6,403	5
	Sandusky	5,796	6
	Toledo	6,494	1
	Youngstown	6,417	4
OK	Oklahoma City	3,725	13
	Tulsa	3,860	13
OR	Eugene	4,726	22
	Medford	5,008	23
	Portland	4,635	23
	Portland	4,109	24
	Salem	4,754	23
PA	Allentown	5,810	9
	Erie	6,451	9
	Harrisburg	5,251	11
	Philadelphia	5,144	14
	Pittsburgh	5,987	5
PA	Pittsburgh	5,053	7
	Reading	4,945	13
	Scranton	6,254	5
	Williamsport	5,934	7
RI	Providence	5,954	9
SC	Charleston	2,033	27
	Charleston	1,794	28
	Columbia	2,484	24
SD	Huron	8,223	-14
	Rapid City	7,345	7
SD	Sioux Falls	7,839	-11
TN	Bristol	4,143	14
	Chattanooga	3,254	18
	Knoxville	3,494	19
	Memphis	3,232	18
	Nashville	3,578	14
TX	Abilene	2,624	20
	Austin	1,711	28
	Dallas	2,363	22
	El Paso	2,700	24
	Houston	1,396	32
	Midland	2,591	21
	San Angelo	2,255	22
	San Antonio	1,546	30
	Waco	2,030	26
	Wichita Falls	2,832	18
UT	Salt Lake City	6,052	8

VT	Burlington	8,269	-7
VA	Lynchburg	4,166	16
	Norfolk	3,421	22
	Richmond	3,865	17
	Roanoke	4,150	16
WA	Olympia	5,236	22
	Seattle-Tacoma	5,145	26
	Seattle	4,424	27
	Spokane	6,655	2
WV	Charleston	4,476	11
	Elkins	5,675	6
	Huntington	4,446	10
	Parkersburg	4,754	11
WI	Green Bay	8,029	-9
	La Crosse	7,589	-9
	Madison	7,863	-7
WI	Milwaukee	7,635	-4
WY	Casper	7,410	-5
	Cheyenne	7,381	-1
	Lander	7,870	-11
	Sheridan	7,680	8

Definitions

Air Change: the amount of air that is required to completely replace the air in the boiler and associated flue passages.

Air, Primary: Air that mixes with the fuel to provide combustion.

Air, Secondary: Air that mixes with the flue gases to dilute the air going outside or to the chimney.

Air Separator: A device located in the supply pipe for a hydronic boiler that removes the entrained air from the water.

Air shutter: A device that controls the airflow to the burner

Backflow Preventer: A device that will limit the backflow of boiler water into the potable water in a building or system.

Barometric Damper: A damper that is installed in the flue piping that will control the excessive draft in a category1 type boiler by introducing boiler room air.

Boiler: A closed vessel that heats water or creates steam

Boiler Design Temperature: It is the outside temperature at which the heating system can still provide heat to the building. It will be one of the coldest temperatures during an average winter.

Boiler, High Pressure: A boiler, which generates steam to pressures above 15 Psig

Boiler, Low Pressure: A boiler, which generates steam to pressures below 15 Psig

Boiler, Hydronic: A boiler, which heats water below the flash point.

Boiler, Cast Iron: A boiler, which uses cast iron as its heat exchanger.

Boiler, Copper: A boiler, which uses copper as its heat exchanger.

Boiler, Steel: A boiler, which uses steel as its heat exchanger.

Boiler, Fire Tube: A steel boiler where the flue gases travel through the tubes inside the boiler.

Boiler, Water Tube: A steel boiler where the flue gases travel around the tubes inside the boiler.

Boiler, Modular: A heating system consisting of several smaller boilers.

Breeching: A conduit that transports the combustion by products from the boiler to the outside or to the chimney. It is also called a flue.

Btu (British Thermal Unit): The amount of heat required to raise one pound of water, one degree F

Btuh: Btu's in one hour

Burner: A mechanical device that mixes air and fuel to provide ignition and combustion of the fuel.

Burner, Atmospheric: A burner that uses natural draft and gas pressure to provide combustion.

Burner, Power: A burner that uses an internal blower to mix the fuel and the air for combustion.

Carbon Dioxide: This is a gas that is produced as a by-product of combustion. It is also referred to as CO_2.

Carbon Monoxide: This deadly gas is odorless and tasteless. It is produced when there the combustion is out of adjustment. It is often referred to as CO.

Compression Tank: A tank that is used in a hydronic system that will absorb the expansion of the water, sometimes called an Compression tank.

Combustion Air: The air that is introduced from the outside that is required for the proper combustion of the fuel.

Combustion Analyzer: A device that measures the flue gas from a boiler and displays the different components. It will also display the efficiency of the boiler.

Condensate: Condensed water because of the removal of latent heat from a gas.

Condensing Boiler: A boiler that is designed to allow the flue temperatures to drop below the dew point temperature.

Control, Operating: A device that starts or stops the burner. This is usually set for a lower temperature or pressure than the Limit Control.

Control, Limit: A device that starts or stops the burner. This is usually set for a higher pressure or temperature than the operating control. In most applications, it has a manual reset feature.

Dew Point Temperature: The temperature at which warm humid air is cooled enough to allow the water vapor to condense into water.

Dirt Leg: A series of nipples and a pipe cap that are installed just before the train to capture any dirt that is in the gas line before it enters the gas train.

Draft: The pressure differential between atmospheric pressure and the pressure in the flue and boiler.

Draft Diverter: An air opening that introduces secondary air to the flue after the main combustion.

Draft, Mechanical: The pressure differential between atmospheric pressure and the pressure in the flue and boiler that is induced because of a fan or blower.

Draft, Natural: The pressure differential between atmospheric pressure and the pressure in the flue and boiler without a fan or blower.

Dual Fuel Burner: A burner, which has two fuel sources that it can use. It is usually natural gas and #2 fuel oil.

Emergency Door Switch: This manual switch is located at all exits from a boiler room that will shut off the boiler in the event that it is engaged.

Expansion Tank: A tank that is used in a hydronic system that will absorb the expansion of the water once it is heated. It is sometimes called a Compression Tank.

Firing Rate: The burning rate of fuel and air in the burner.

Firing Rate Control: A control that senses the temperature or pressure of the heating system. It will regulate the burner between low and high fire to meet the desired set point. It is sometimes called the modulating control.

Flue: A conduit that transports the combustion by products from the boiler to the outside or to the chimney. It is also called a breeching.

Flue Gases: These byproducts of combustion are produced by the burner. They will be vented from the boiler with a flue.

Fuel Train: A series of components, including gas pressure regulator and gas valves, that are located in the gas piping directly attached to the burner. This is also called a gas train.

Gas Pressure Regulator: A device that controls the gas pressure supplied to the burner.

Gas Pressure Switch: A safety device that senses the available gas pressure and will shut the boiler off in the event that the pressure is outside of the setting. There are usually two types of gas pressure switches on a boiler. The High Gas Pressure Switch that is located in the gas train downstream of the gas pressure regulator and the electric gas valves. It will shut the boiler off if the gas pressure is higher than the setting. The Low Gas Pressure Switch is located downstream of the main gas pressure regulator. It will shut the boiler off if the gas pressure is below the setting.

Heat, Latent: The amount of heat required to cause a change of state.

Heat, Sensible: The amount of heat required to cause a change in temperature

Heating Medium: The material that the boiler heats. It could be steam, water or some other type of fluid.

High Fire: This is the highest design firing rate of the burner. It is the 100% firing rate.

Hydronic System: A heating system that uses water as the heating medium instead of steam.

Lag Boiler: The boiler that is not the first boiler to start when there is a call for heat.

Lead Boiler: The boiler that is the first boiler to start on a call for heat.

Life Cycle Cost: This is the amount of money that the system costs the owner over the estimated life of the unit. It will include fuel and repair costs as well estimated repair parts.

Lockout: A safety shutdown that requires a manual reset of the control or safety device.

Low Fire: This is the lowest design firing rate of the burner.

Low Fire Start: This switch verifies that the burner is in the "Low Fire" position before opening fuel valves.

Low High Low Fire: A burner that starts at low fire and then goes to high fire if there is still a call for heat. As the temperature or pressure gets close to the set point on the firing rate control, the burner will drop to low fire.

Low High Off Fire: A burner that starts at low fire and then goes to high fire if there is still a call for heat. The burner will stay at high fire until the call for heat has ceased.

Low Water Cutoff: A device that senses the water level inside the boiler and will shut down the burner if the water level drops to an unsafe level.

Modulating Burner: A burner that will operate at any position from low to high fire to meet the demands of the modulating control.

Modulating Control: A control that senses the heating medium and will send a signal to the burner that will set the burner at any position from low to high fire.

Non-Condensing Boiler: A boiler that is designed to keep the flue temperatures above the dew point temperature.

Pilot, Continuous: It is a pilot flame that burns all the time, regardless of whether the burner is firing.

Pilot, Intermittent. It is a pilot that lights when there is a call for heat. The pilot will stay light during the entire time that the main burner is firing.

Pilot, Interrupted: It is a pilot that lights when there is a call for heat. The pilot will shut off once the main flame is established.

Pot Feeder: A device that is used to introduce water treatment into a heating system.
PPM: Part per Million.

Prepurge: On a call for heat, the burner blower starts to purge the boiler combustion chamber and flue passages of any unburnt fuels. It will operate for a duration long enough to provide several air changes inside the boiler.

Proprietary Parts: parts that are only available from the manufacturer or have limited distribution.

Post purge: The burner blower will operate for a time after the call for heat has been satisfied to purge any unburnt fuel.

Relief Valve: A valve located on a boiler that will relieve the internal boiler pressure if the pressure rises to the rating of the relief valve.

Reset Control: A control that will lower the supply temperature in a hydronic system as the outside temperature increases.

Sidewall Venting: Boiler flue that is piped to the sidewall of the building rather than a chimney or stack.

Spill Switch: A device that is located by a draft diverter or a barometric damper that senses rollout of the flue gases and shuts off the burner.

Index